Multivariate Statistical Methods

· A primer ·

THIRD EDITION

Bryan F. J. Manly

Western EcoSystems Technology, Inc.
Laramie, Wyoming, USA

CHAPMAN & HALL/CRC

A CRC Press Company
Boca Raton   London   New York   Washington, D.C.

Library of Congress Cataloging-in-Publication Data

Manly, Bryan F. J., 1944–
    Multivariate statistical methods : a primer / Bryan F.J. Manly. — 3rd ed.
        p. cm.
    Includes bibliographical references and index.
    ISBN 1-58488-414-2 (alk. paper)
        1. Multivariate analysis. I. Title.

QA278.M35 2004
519.5'35—dc22

                                                                        2004045489

Visit the CRC Press Web site at www.crcpress.com

© 2005 by Chapman & Hall/CRC

No claim to original U.S. Government works
International Standard Book Number 1-58488-414-2
Library of Congress Card Number 2004045489
Printed in the United States of America   4 5 6 7 8 9 0
Printed on acid-free paper

A journey of a thousand miles begins with a single step

Lao Tsu

Contents

Chapter 1   The material of multivariate analysis 1
1.1   Examples of multivariate data 1
1.2   Preview of multivariate methods 12
1.3   The multivariate normal distribution 14
1.4   Computer programs 15
1.5   Graphical methods 15
1.6   Chapter summary 16
References 16

Chapter 2   Matrix algebra 17
2.1   The need for matrix algebra 17
2.2   Matrices and vectors 17
2.3   Operations on matrices 19
2.4   Matrix inversion 21
2.5   Quadratic forms 22
2.6   Eigenvalues and eigenvectors 22
2.7   Vectors of means and covariance matrices 23
2.8   Further reading 25
2.9   Chapter summary 25
References 26

Chapter 3   Displaying multivariate data 27
3.1   The problem of displaying many variables in two dimensions 27
3.2   Plotting index variables 27
3.3   The draftsman's plot 29
3.4   The representation of individual data points 30
3.5   Profiles of variables 32
3.6   Discussion and further reading 33
3.7   Chapter summary 34
References 34

Chapter 4   Tests of significance with multivariate data 35
4.1   Simultaneous tests on several variables 35
4.2   Comparison of mean values for two samples: the single variable
        case 35

4.3   Comparison of mean values for two samples: the multivariate
      case 37
4.4   Multivariate versus univariate tests 41
4.5   Comparison of variation for two samples: the single-variable
      case 42
4.6   Comparison of variation for two samples: the multivariate case 42
4.7   Comparison of means for several samples 46
4.8   Comparison of variation for several samples 49
4.9   Computer programs 54
4.10 Chapter summary 54
Exercise 55
References 57

Chapter 5   Measuring and testing multivariate distances **59**
5.1   Multivariate distances 59
5.2   Distances between individual observations 59
5.3   Distances between populations and samples 62
5.4   Distances based on proportions 67
5.5   Presence–absence data 68
5.6   The Mantel randomization test 69
5.7   Computer programs 72
5.8   Discussion and further reading 72
5.9   Chapter summary 73
Exercise 74
References 74

Chapter 6   Principal components analysis **75**
6.1   Definition of principal components 75
6.2   Procedure for a principal components analysis 76
6.3   Computer programs 84
6.4   Further reading 85
6.5   Chapter summary 85
Exercises 87
References 90

Chapter 7   Factor analysis **91**
7.1   The factor analysis model 91
7.2   Procedure for a factor analysis 93
7.3   Principal components factor analysis 95
7.4   Using a factor analysis program to do principal components
      analysis 97
7.5   Options in analyses 100
7.6   The value of factor analysis 101
7.7   Computer programs 101
7.8   Discussion and further reading 102
7.9   Chapter summary 102

Exercise 103
References 103

Chapter 8   Discriminant function analysis **105**
8.1   The problem of separating groups 105
8.2   Discrimination using Mahalanobis distances 105
8.3   Canonical discriminant functions 107
8.4   Tests of significance 108
8.5   Assumptions 109
8.6   Allowing for prior probabilities of group membership ... 114
8.7   Stepwise discriminant function analysis 114
8.8   Jackknife classification of individuals 116
8.9   Assigning of ungrouped individuals to groups 116
8.10 Logistic regression 117
8.11 Computer programs 122
8.12 Discussion and further reading 122
8.13 Chapter summary 123
Exercises 124
References 124

Chapter 9   Cluster analysis **125**
9.1   Uses of cluster analysis 125
9.2   Types of cluster analysis 125
9.3   Hierarchic methods 127
9.4   Problems of cluster analysis 129
9.5   Measures of distance 129
9.6   Principal components analysis with cluster analysis ... 130
9.7   Computer programs 134
9.8   Discussion and further reading 135
9.9   Chapter summary 136
Exercises 137
References 141

Chapter 10   Canonical correlation analysis **143**
10.1 Generalizing a multiple regression analysis 143
10.2 Procedure for a canonical correlation analysis 145
10.3 Tests of significance 146
10.4 Interpreting canonical variates 148
10.5 Computer programs 158
10.6 Further reading 158
10.7 Chapter summary 159
Exercise 159
References 161

Chapter 11   Multidimensional scaling **163**
11.1 Constructing a map from a distance matrix 163

11.2 Procedure for multidimensional scaling...............................165
11.3 Computer programs...172
11.4 Further reading 174
11.5 Chapter summary..174
Exercise...175
References..175

Chapter 12   Ordination **177**
12.1 The ordination problem..177
12.2 Principal components analysis 178
12.3 Principal coordinates analysis 181
12.4 Multidimensional scaling 189
12.5 Correspondence analysis 191
12.6 Comparison of ordination methods 196
12.7 Computer programs 197
12.8 Further reading 197
12.9 Chapter summary 198
Exercise 198
References..198

Chapter 13   Epilogue **201**
13.1 The next step 201
13.2 Some general reminders 201
13.3 Missing values..202
References..203

Appendix   Computer packages for multivariate analyses....................**205**
References..207

Author Index...**209**

Subject Index **211**

Preface

The purpose of this book is to introduce multivariate statistical methods to non-mathematicians. It is not intended to be a comprehensive textbook. Rather, the intention is to keep the details to a minimum while serving as a practical guide that illustrates the possibilities of multivariate statistical analysis. In other words, it is a book to "get you going" in a particular area of statistical methods.

It is assumed that readers have a working knowledge of elementary statistics, including tests of significance using normal-, t-, chi-squared, and F-distributions; analysis of variance; and linear regression. The material covered in a typical first-year university course in statistics should be quite adequate in this respect. Some facility with algebra is also required to follow the equations in certain parts of the text. Understanding the theory of multivariate methods requires some matrix algebra. However, the amount needed is not great if some details are accepted on faith. Matrix algebra is summarized in Chapter 2, and anyone that masters this chapter will have a reasonable competency in this area.

One of the reasons why multivariate methods are being used so often these days is the ready availability of computer packages to do the calculations. Indeed, access to suitable computer software is essential if the methods are to be used. However, the details of the use of computer packages are not stressed in this book because there are so many of these packages available. It would be impossible to discuss them all, and it would be too restrictive to concentrate on one or two of them. The approach taken here is to mention which package was used for a particular example when this is appropriate. In addition, the Appendix gives information about some of the packages in terms of what analyses are available and how easy the programs are to use for someone who is relatively inexperienced at carrying out multivariate analyses.

To some extent, the chapters can be read independently of each other. The first five are preliminary reading, focusing mainly on general aspects of multivariate data rather than specific techniques. Chapter 1 introduces data for several examples that are used to illustrate the application of analytical methods throughout the book. Chapter 2 covers matrix algebra, and Chapter 3 discusses various graphical techniques. Chapter 4 discusses tests of significance, and Chapter 5 addresses the measurement of relative "distances" between objects based on variables measured on those objects. These chapters should be reviewed before Chapters 6 to 12, which cover the most important multivariate procedures in current use. The final Chapter 13 contains some general comments about the analysis of multivariate data.

The chapters in this third edition of the book are the same as those in the second edition. The changes that have been made for the new edition are the updating of references, some new examples, some examples carried out using newer computer software, and changes in the text to reflect new ideas about multivariate analyses.

In making changes, I have continually kept in mind the original intention of the book, which was that it should be as short as possible and attempt to do no more than take readers to the stage where they can begin to use multivariate methods in an intelligent manner.

I am indebted to many people for commenting on the text of the three editions of the book and for pointing out various errors. Particularly, I thank Earl Bardsley, John Harraway, and Liliana Gonzalez for their help in this respect. Any errors that remain are my responsibility alone.

I would like to express my appreciation to the Department of Mathematics and Statistics at the University of Otago in New Zealand for hosting me as a visitor twice in 2003, first in May and June, and later in November and December. The excellent university library was particularly important for my final updating of references.

In conclusion, I wish to thank the staff of Chapman and Hall and of CRC for their work over the years in promoting the book and encouraging me to produce the second and third editions.

Bryan F.J. Manly
Laramie, Wyoming

chapter one

The material of multivariate analysis

1.1 Examples of multivariate data

The statistical methods that are described in elementary texts are mostly univariate methods because they are only concerned with analyzing variation in a single random variable. On the other hand, the whole point of a multivariate analysis is to consider several related variables simultaneously, with each one being considered to be equally important, at least initially. The potential value of this more general approach can be seen by considering a few examples.

Example 1.1 Storm survival of sparrows

After a severe storm on 1 February 1898, a number of moribund sparrows were taken to Hermon Bumpus' biological laboratory at Brown University in Rhode Island. Subsequently about half of the birds died, and Bumpus saw this as an opportunity to see whether he could find any support for Charles Darwin's theory of natural selection. To this end, he made eight morphological measurements on each bird, and also weighed them. The results for five of the measurements are shown in Table 1.1, for females only.

From the data that he obtained, Bumpus (1898) concluded that "the birds which perished, perished not through accident, but because they were physically disqualified, and that the birds which survived, survived because they possessed certain physical characters." Specifically, he found that the survivors "are shorter and weigh less ... have longer wing bones, longer legs, longer sternums and greater brain capacity" than the nonsurvivors. He also concluded that "the process of selective elimination is most severe with extremely variable individuals, no matter in which direction the variation may occur. It is quite as dangerous to be above a certain standard of organic excellence as it is to be conspicuously below the standard." This was saying that stabilizing selection occurred, so that individuals with measurements

Table 1.1 Body Measurements of Female Sparrows

Bird	X_1 (mm)	X_2 (mm)	X_3 (mm)	X_4 (mm)	X_5 (mm)
1	156	245	31.6	18.5	20.5
2	154	240	30.4	17.9	19.6
3	153	240	31.0	18.4	20.6
4	153	236	30.9	17.7	20.2
5	155	243	31.5	18.6	20.3
6	163	247	32.0	19.0	20.9
7	157	238	30.9	18.4	20.2
8	155	239	32.8	18.6	21.2
9	164	248	32.7	19.1	21.1
10	158	238	31.0	18.8	22.0
11	158	240	31.3	18.6	22.0
12	160	244	31.1	18.6	20.5
13	161	246	32.3	19.3	21.8
14	157	245	32.0	19.1	20.0
15	157	235	31.5	18.1	19.8
16	156	237	30.9	18.0	20.3
17	158	244	31.4	18.5	21.6
18	153	238	30.5	18.2	20.9
19	155	236	30.3	18.5	20.1
20	163	246	32.5	18.6	21.9
21	159	236	31.5	18.0	21.5
22	155	240	31.4	18.0	20.7
23	156	240	31.5	18.2	20.6
24	160	242	32.6	18.8	21.7
25	152	232	30.3	17.2	19.8
26	160	250	31.7	18.8	22.5
27	155	237	31.0	18.5	20.0
28	157	245	32.2	19.5	21.4
29	165	245	33.1	19.8	22.7
30	153	231	30.1	17.3	19.8
31	162	239	30.3	18.0	23.1
32	162	243	31.6	18.8	21.3
33	159	245	31.8	18.5	21.7
34	159	247	30.9	18.1	19.0
35	155	243	30.9	18.5	21.3
36	162	252	31.9	19.1	22.2
37	152	230	30.4	17.3	18.6
38	159	242	30.8	18.2	20.5
39	155	238	31.2	17.9	19.3
40	163	249	33.4	19.5	22.8
41	163	242	31.0	18.1	20.7
42	156	237	31.7	18.2	20.3
43	159	238	31.5	18.4	20.3
44	161	245	32.1	19.1	20.8
45	155	235	30.7	17.7	19.6

(continued)

Table 1.1 (continued)   Body Measurements of Female Sparrows

Bird	X_1 (mm)	X_2 (mm)	X_3 (mm)	X_4 (mm)	X_5 (mm)
46	162	247	31.9	19.1	20.4
47	153	237	30.6	18.6	20.4
48	162	245	32.5	18.5	21.1
49	164	248	32.3	18.8	20.9

Note: X_1 = total length, X_2 = alar extent, X_3 = length of beak and head, X_4 = length of humerus, and X_5 = length of keel of sternum. Birds 1 to 21 survived, and birds 22 to 49 died. The data source is Bumpus (1898), who measured in inches and millimeters.

Source: Adapted from Bumpus, H.C. (1898), *Biological Lectures*, 11th Lecture, Marine Biology Laboratory, Woods Hole, MA, pp. 209–226.

close to the average survived better than individuals with measurements far from the average.

In fact, the development of multivariate statistical methods had hardly begun in 1898 when Bumpus was writing. The correlation coefficient as a measure of the relationship between two variables was devised by Francis Galton in 1877. However, it was another 56 years before Harold Hotelling described a practical method for carrying out a principal components analysis, which is one of the simplest multivariate analyses that can usefully be applied to Bumpus' data. Bumpus did not even calculate standard deviations. Nevertheless, his methods of analysis were sensible. Many authors have reanalyzed his data and, in general, have confirmed his conclusions.

Taking the data as an example for illustrating multivariate methods, several interesting questions arise. In particular:

1. How are the various measurements related? For example, does a large value for one of the variables tend to occur with large values for the other variables?
2. Do the survivors and nonsurvivors have statistically significant differences for their mean values of the variables?
3. Do the survivors and nonsurvivors show similar amounts of variation for the variables?
4. If the survivors and nonsurvivors do differ in terms of the distributions of the variables, then is it possible to construct some function of these variables that separates the two groups? It would then be convenient if large values of the function tended to occur with the survivors, as the function would then apparently be an index of the Darwinian fitness of the sparrows.

Example 1.2   Egyptian skulls

For a second example, consider the data shown in Table 1.2 for measurements made on male skulls from the area of Thebes in Egypt. There are five samples of 30 skulls each from the early predynastic period (*circa* 4000 B.C.), the late

Table 1.2 Measurement on Male Egyptian Skulls (mm)

Skull	Early Predynastic				Late Predynastic				12th and 13th Dynasties				Ptolemaic Period				Roman Period			
	X_1	X_2	X_3	X_4	X_1	X_2	X_3	X_4	X_1	X_2	X_3	X_4	X_1	X_2	X_3	X_4	X_1	X_2	X_3	X_4
1	131	138	89	49	124	138	101	48	137	141	96	52	137	134	107	54	137	123	91	50
2	125	131	92	48	133	134	97	48	129	133	93	47	141	128	95	53	136	131	95	49
3	131	132	99	50	138	134	98	45	132	138	87	48	141	130	87	49	128	126	91	57
4	119	132	96	44	148	129	104	51	130	134	106	50	135	131	99	51	130	134	92	52
5	136	143	100	54	126	124	95	45	134	134	96	45	133	120	91	46	138	127	86	47
6	138	137	89	56	135	136	98	52	140	133	98	50	131	135	90	50	126	138	101	52
7	139	130	108	48	132	145	100	54	138	138	95	47	140	137	94	60	136	138	97	58
8	125	136	93	48	133	130	102	48	136	145	99	55	139	130	90	48	126	126	92	45
9	131	134	102	51	131	134	96	50	136	131	92	46	140	134	90	51	132	132	99	55
10	134	134	99	51	133	125	94	46	126	136	95	56	138	140	100	52	139	135	92	54
11	129	138	95	50	133	136	103	53	137	129	100	53	132	133	90	53	143	120	95	51
12	134	121	95	53	131	139	98	51	137	139	97	50	134	134	97	54	141	136	101	54
13	126	129	109	51	131	136	99	56	136	126	101	50	135	135	99	50	135	135	95	56
14	132	136	100	50	138	134	98	49	137	133	90	49	133	136	95	52	137	134	93	53
15	141	140	100	51	130	136	104	53	129	142	104	47	136	130	99	55	142	135	96	52
16	131	134	97	54	131	128	98	45	135	138	102	55	134	137	93	52	139	134	95	47

	X_1	X_2	X_3	X_4	X_1	X_2	X_3	X_4	X_1	X_2	X_3	X_4	X_1	X_2	X_3	X_4	X_1	X_2	X_3	X_4
17	135	137	103	50	138	129	107	53	129	135	92	50	131	141	99	55	138	125	99	51
18	132	133	93	53	123	131	101	51	134	125	90	60	129	135	95	47	137	135	96	54
19	139	136	96	50	130	129	105	47	138	134	96	51	136	128	93	54	133	125	92	50
20	132	131	101	49	134	130	93	54	136	135	94	53	131	125	88	48	145	129	89	47
21	126	133	102	51	137	136	106	49	132	130	91	52	139	130	94	53	138	136	92	46
22	135	135	103	47	126	131	100	48	133	131	100	50	144	124	86	50	131	129	97	44
23	134	124	93	53	135	136	97	52	138	137	94	51	141	131	97	53	143	126	88	54
24	128	134	103	50	129	126	91	50	130	127	99	45	130	131	98	53	134	124	91	55
25	130	130	104	49	134	139	101	49	136	133	91	49	133	128	92	51	132	127	97	52
26	138	135	100	55	131	134	90	53	134	123	95	52	138	126	97	54	137	125	85	57
27	128	132	93	53	132	130	104	50	136	137	101	54	131	142	95	53	129	128	81	52
28	127	129	106	48	130	132	93	52	133	131	96	49	136	138	94	55	140	135	103	48
29	131	136	114	54	135	132	98	54	138	133	100	55	132	136	92	52	147	129	87	48
30	124	138	101	46	130	128	101	51	138	133	91	46	135	130	100	51	136	133	97	51

Note: X_1 = maximum breadth, X_2 = basibregmatic height, X_3 = basialveolar length, X_4 = nasal height.

Source: From Thomson, A. and Randall-Maciver, R. (1905), *Ancient Races of the Thebaid*, Oxford University Press, Oxford, U.K.

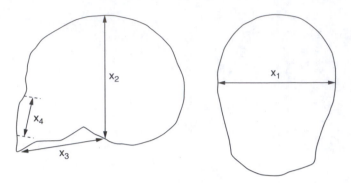

Figure 1.1 Four measurements made on Egyptian male skulls.

predynastic period (*circa* 3300 B.C.), the 12th and 13th dynasties (*circa* 1850 B.C.), the Ptolemaic period (*circa* 200 B.C.), and the Roman period (*circa* A.D. 150). Four measurements are available on each skull, as illustrated in Figure 1.1.

For this example, some interesting questions are:

1. How are the four measurements related?
2. Are there statistically significant differences in the sample means for the variables, and if so, do these differences reflect gradual changes over time in the shape and size of skulls?
3. Are there significant differences in the sample standard deviations for the variables, and, if so, do these differences reflect gradual changes over time in the amount of variation?
4. Is it possible to construct a function of the four variables that, in some sense, describes the changes over time?

These questions are, of course, rather similar to the ones suggested for Example 1.1.

It will be seen later that there are differences between the five samples that can be explained partly as time trends. It must be said, however, that the reasons for the apparent changes are unknown. Migration of other races into the region may well have been the most important factor.

Example 1.3 Distribution of a butterfly

A study of 16 colonies of the butterfly *Euphydryas editha* in California and Oregon produced the data shown in Table 1.3. Here there are four environmental variables (altitude, annual precipitation, and the minimum and maximum temperatures) and six genetic variables (percentage frequencies for different Pgi genes as determined by the technique of electrophoresis). For the purposes of this example, there is no need to go into the details of how the gene frequencies were determined, and, strictly speaking, they are not exactly gene frequencies anyway. It is sufficient to say that the frequencies

Table 1.3 Environmental Variables and Phosphoglucose-Isomerase (Pgi) Gene Frequencies for Colonies of the Butterfly *Euphydryas editha* in California and Oregon[a]

Colony	Altitude (ft)	Annual Precipitation (in.)	Temperature (°F) Maximum	Minimum	Frequencies of Pgi Mobility Genes (%)[b] 0.4	0.6	0.8	1	1.16	1.3
SS	500	43	98	17	0	3	22	57	17	1
SB	808	20	92	32	0	16	20	38	13	13
WSB	570	28	98	26	0	6	28	46	17	3
JRC	550	28	98	26	0	4	19	47	27	3
JRH	550	28	98	26	0	1	8	50	35	6
SJ	380	15	99	28	0	2	19	44	32	3
CR	930	21	99	28	0	0	15	50	27	8
UO	650	10	101	27	10	21	40	25	4	0
LO	600	10	101	27	14	26	32	28	0	0
DP	1,500	19	99	23	0	1	6	80	12	1
PZ	1,750	22	101	27	1	4	34	33	22	6
MC	2,000	58	100	18	0	7	14	66	13	0
IF	2,500	34	102	16	0	9	15	47	21	8
AF	2,000	21	105	20	3	7	17	32	27	14
GH	7,850	42	84	5	0	5	7	84	4	0
GL	10,500	50	81	-12	0	3	1	92	4	0

[a] The data source was McKechnie et al. (1975), with the environmental variables rounded to integers for simplicity. The original data were for 21 colonies, but for the present example, five colonies with small samples for the estimation of gene frequencies have been excluded to make all estimates about equally reliable.

[b] The numbers 0.40, 0.60, etc. represent different genetic types of Pgi so that the frequencies for a colony (adding to 100%) show the frequencies of the different types for the *E. editha* at that location.

Source: Adapted from McKechnie, S.W. et al. (1975), *Genetics*, 81: 571–594.

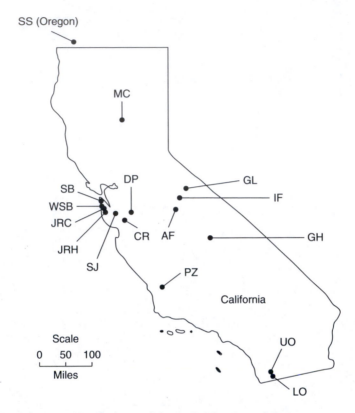

Figure 1.2 Colonies of *Euphydryas editha* in California and Oregon.

describe the genetic distribution of the butterfly to some extent. Figure 1.2 shows the geographical locations of the colonies.

In this example, questions that can be asked include:

1. Are the Pgi frequencies similar for colonies that are close in space?
2. To what extent, if any, are the Pgi frequencies related to the environmental variables?

These are important questions in trying to decide how the Pgi frequencies are determined. If the genetic composition of the colonies was largely determined by past and present migration, then gene frequencies will tend to be similar for colonies that are close in space, although they may show little relationship with the environmental variables. On the other hand, if it is the environment that is most important, then this should show up in relationships between the gene frequencies and the environmental variables (assuming that the right variables have been measured), but close colonies will only have similar gene frequencies if they have similar environments. Obviously, colonies that are close in space will usually have similar environments, so it may be difficult to reach a clear conclusion on this matter.

Table 1.4 Mean Mandible Measurements for Seven Canine Groups

Group	X_1 (mm)	X_2 (mm)	X_3 (mm)	X_4 (mm)	X_5 (mm)	X_6 (mm)
Modern dog	9.7	21.0	19.4	7.7	32.0	36.5
Golden jackal	8.1	16.7	18.3	7.0	30.3	32.9
Chinese wolf	13.5	27.3	26.8	10.6	41.9	48.1
Indian wolf	11.5	24.3	24.5	9.3	40.0	44.6
Cuon	10.7	23.5	21.4	8.5	28.8	37.6
Dingo	9.6	22.6	21.1	8.3	34.4	43.1
Prehistoric dog	10.3	22.1	19.1	8.1	32.2	35.0

Note: X_1 = breadth of mandible; X_2 = height of mandible below the first molar; X_3 = length of the first molar; X_4 = breadth of the first molar; X_5 = length from first to third molar, inclusive; and X_6 = length from first to fourth premolar, inclusive.

Source: Adapted from Higham, C.F.W. et al. (1980), *J. Archaeological Sci., 7*, 149–165.

Example 1.4   Prehistoric dogs from Thailand

Excavations of prehistoric sites in northeast Thailand have produced a collection of canid (dog) bones covering a period from about 3500 B.C. to the present. However, the origin of the prehistoric dog is not certain. It could descend from the golden jackal (*Canis aureus*) or the wolf, but the wolf is not native to Thailand. The nearest indigenous sources are western China (*Canis lupus chanco*) or the Indian subcontinent (*Canis lupus pallides*).

In order to try to clarify the ancestors of the prehistoric dogs, mandible (lower jaw) measurements were made on the available specimens. These were then compared with the same measurements made on the golden jackal, the Chinese wolf, and the Indian wolf. The comparisons were also extended to include the dingo, which may have its origins in India; the cuon (*Cuon alpinus*), which is indigenous to southeast Asia; and modern village dogs from Thailand.

Table 1.4 gives mean values for the six mandible measurements for specimens from all seven groups. The main question here is what the measurements suggest about the relationships between the groups and, in particular, how the prehistoric dog seems to relate to the other groups.

Example 1.5   Employment in European countries

Finally, as a contrast to the previous biological examples, consider the data in Table 1.5. This shows the percentages of the labor force in nine different types of industry for 30 European countries. In this case, multivariate methods may be useful in isolating groups of countries with similar employment patterns, and in generally aiding the understanding of the relationships between the countries. Differences between countries that are related to political grouping (EU, the European Union; EFTA, the European Free Trade Area; the Eastern European countries; and other countries) may be of particular interest.

Table 1.5 Percentages of the Workforce Employed in Nine Different Industry Groups in 30 Countries in Europe

Country	Group	AGR	MIN	MAN	PS	CON	SER	FIN	SPS	TC
Belgium	EU	2.6	0.2	20.8	0.8	6.3	16.9	8.7	36.9	6.8
Denmark	EU	5.6	0.1	20.4	0.7	6.4	14.5	9.1	36.3	7.0
France	EU	5.1	0.3	20.2	0.9	7.1	16.7	10.2	33.1	6.4
Germany	EU	3.2	0.7	24.8	1.0	9.4	17.2	9.6	28.4	5.6
Greece	EU	22.2	0.5	19.2	1.0	6.8	18.2	5.3	19.8	6.9
Ireland	EU	13.8	0.6	19.8	1.2	7.1	17.8	8.4	25.5	5.8
Italy	EU	8.4	1.1	21.9	0.0	9.1	21.6	4.6	28.0	5.3
Luxembourg	EU	3.3	0.1	19.6	0.7	9.9	21.2	8.7	29.6	6.8
Netherlands	EU	4.2	0.1	19.2	0.7	0.6	18.5	11.5	38.3	6.8
Portugal	EU	11.5	0.5	23.6	0.7	8.2	19.8	6.3	24.6	4.8
Spain	EU	9.9	0.5	21.1	0.6	9.5	20.1	5.9	26.7	5.8
U.K.	EU	2.2	0.7	21.3	1.2	7.0	20.2	12.4	28.4	6.5
Austria	EFTA	7.4	0.3	26.9	1.2	8.5	19.1	6.7	23.3	6.4
Finland	EFTA	8.5	0.2	19.3	1.2	6.8	14.6	8.6	33.2	7.5
Iceland	EFTA	10.5	0.0	18.7	0.9	10.0	14.5	8.0	30.7	6.7
Norway	EFTA	5.8	1.1	14.6	1.1	6.5	17.6	7.6	37.5	8.1
Sweden	EFTA	3.2	0.3	19.0	0.8	6.4	14.2	9.4	39.5	7.2
Switzerland	EFTA	5.6	0.0	24.7	0.0	9.2	20.5	10.7	23.1	6.2

Country	Region	AGR	MIN	MAN	PS	CON	SER	FIN	SPS	TC
Albania	Eastern	55.5	19.4	0.0	0.0	3.4	3.3	15.3	0.0	3.0
Bulgaria	Eastern	19.0	0.0	35.0	0.0	6.7	9.4	1.5	20.9	7.5
Czech/Slovak Republics	Eastern	12.8	37.3	0.0	0.0	8.4	10.2	1.6	22.9	6.9
Hungary	Eastern	15.3	28.9	0.0	0.0	6.4	13.3	0.0	27.3	8.8
Poland	Eastern	23.6	3.9	24.1	0.9	6.3	10.3	1.3	24.5	5.2
Romania	Eastern	22.0	2.6	37.9	2.0	5.8	6.9	0.6	15.3	6.8
USSR (former)	Eastern	18.5	0.0	28.8	0.0	10.2	7.9	0.6	25.6	8.4
Yugoslavia (former)	Eastern	5.0	2.2	38.7	2.2	8.1	13.8	3.1	19.1	7.8
Cyprus	Other	13.5	0.3	19.0	0.5	9.1	23.7	6.7	21.2	6.0
Gibraltar	Other	0.0	0.0	6.8	2.0	16.9	24.5	10.8	34.0	5.0
Malta	Other	2.6	0.6	27.9	1.5	4.6	10.2	3.9	41.6	7.2
Turkey	Other	44.8	0.9	15.3	0.2	5.2	12.4	2.4	14.5	4.4

Note: AGR, agriculture, forestry, and fishing; MIN, mining and quarrying; MAN, manufacturing; PS, power and water supplies; CON, construction; SER, services; FIN, finance; SPS, social and personal services; TC, transport and communications. The data for the individual countries are for various years from 1989 to 1995. Data from Euromonitor (1995), except for Germany and the U.K., where more reasonable values were obtained from the United Nations Statistical Yearbook (2000).

Source: Adapted from Euromonitor (1995), *European Marketing Data and Statistics*, Euromonitor Publications, London; and from United Nations (2000), *Statistical Yearbook*, 44th Issue, U.N. Department of Social Affairs, New York.

1.2 *Preview of multivariate methods*

The five examples just considered are typical of the raw material for multi-variate statistical methods. In all cases, there are several variables of interest, and these are clearly not independent of each other. At this point, it is useful to give a brief preview of what is to come in the chapters that follow in relationship to these examples.

Principal components analysis is designed to reduce the number of variables that need to be considered to a small number of indices (called the principal components) that are linear combinations of the original variables. For example, much of the variation in the body measurements of sparrows (X_1 to X_5) shown in Table 1.1 will be related to the general size of the birds, and the total

$$I_1 = X_1 + X_2 + X_3 + X_4 + X_5$$

should measure this aspect of the data quite well. This accounts for one dimension of the data. Another index is

$$I_2 = X_1 + X_2 + X_3 - X_4 - X_5$$

which is a contrast between the first three measurements and the last two. This reflects another dimension of the data. Principal components analysis provides an objective way of finding indices of this type so that the variation in the data can be accounted for as concisely as possible. It may well turn out that two or three principal components provide a good summary of all the original variables. Consideration of the values of the principal components instead of the values of the original variables may then make it much easier to understand what the data have to say. In short, principal components analysis is a means of simplifying data by reducing the number of variables.

Factor analysis also attempts to account for the variation in a number of original variables using a smaller number of index variables or factors. It is assumed that each original variable can be expressed as a linear combination of these factors, plus a residual term that reflects the extent to which the variable is independent of the other variables. For example, a two-factor model for the sparrow data assumes that

$$X_1 = a_{11}F_1 + a_{12}F_2 + e_1$$
$$X_2 = a_{21}F_1 + a_{22}F_2 + e_2$$
$$X_3 = a_{31}F_1 + a_{32}F_2 + e_3$$
$$X_4 = a_{41}F_1 + a_{42}F_2 + e_4$$
$$X_5 = a_{51}F_1 + a_{52}F_2 + e_5$$

where the a_{ij} values are constants, F_1 and F_2 are the factors, and e_i represents the variation in X_i that is independent of the variation in the other X variables. Here F_1 might be the factor of size. In that case, the coefficients a_{11}, a_{21}, a_{31}, a_{41}, and a_{51} would all be positive, reflecting the fact that some birds tend to be large and some birds tend to be small on all body measurements. The second factor F_2 might then measure an aspect of the shape of birds, with some positive coefficients and some negative coefficients. If this two-factor model fitted the data well, then it would provide a relatively straightforward description of the relationship between the five body measurements being considered.

One type of factor analysis starts by taking the first few principal components as the factors in the data being considered. These initial factors are then modified by a special transformation process called factor rotation in order to make them easier to interpret. Other methods for finding initial factors are also used. A rotation to simpler factors is almost always done.

Discriminant function analysis is concerned with the problem of seeing whether it is possible to separate different groups on the basis of the available measurements. This could be used, for example, to see how well surviving and nonsurviving sparrows can be separated using their body measurements (Example 1.1), or how skulls from different epochs can be separated, again using size measurements (Example 1.2). Like principal components analysis, discriminant function analysis is based on the idea of finding suitable linear combinations of the original variables to achieve the intended aim.

Cluster analysis is concerned with the identification of groups of similar objects. There is not much point in doing this type of analysis with data like those of Example 1.1 and 1.2, as the groups (survivors/nonsurvivors and epochs) are already known. However, in Example 1.3 there might be some interest in grouping colonies on the basis of environmental variables or Pgi frequencies, while in Example 1.4 the main point of interest is in the similarity between prehistoric Thai dogs and other animals. Likewise, in Example 1.5 the European countries can possibly be grouped in terms of their similarity in employment patterns.

With *canonical correlation*, the variables (not the objects) are divided into two groups, and interest centers on the relationship between these. Thus in Example 1.3, the first four variables are related to the environment, while the remaining six variables reflect the genetic distribution at the different colonies of *Euphydryas editha*. Finding what relationships, if any, exist between these two groups of variables is of considerable biological interest.

Multidimensional scaling begins with data on some measure of the distances apart of a number of objects. From these distances, a map is then constructed showing how the objects are related. This is a useful facility, as it is often possible to measure how far apart pairs of objects are without having any idea of how the objects are related in a geometric sense. Thus in Example 1.4, there are ways of measuring the distances between modern dogs and golden jackals, modern dogs and Chinese wolves, etc. Considering each pair of animal groups gives 21 distances altogether, and from these distances multidimensional scaling can be used to produce a type of map of the relationships between the groups. With a one-dimensional map, the groups are placed along a straight line. With a two-dimensional map, they are represented by points on a plane. With a three-dimensional map, they are represented by points within a cube. Four-dimensional and higher solutions are also possible, although these have limited use because they cannot be visualized in a simple way. The value of a one-, two-, or three-dimensional map is clear for Example 1.4, as such a map would immediately show which groups prehistoric dogs are most similar to. Hence multidimensional scaling may be a useful alternative to cluster analysis in this case. A map of European countries based on their employment patterns might also be of interest in Example 1.5.

Principal components analysis and multidimensional scaling are sometimes referred to as methods for *ordination*. That is to say, they are methods for producing axes against which a set of objects of interest can be plotted. Other methods of ordination are also available.

Principal coordinates analysis is like a type of principal components analysis that starts off with information on the extent to which the pairs of objects are different in a set of objects, instead of the values for measurements on the objects. As such, it is intended to do the same as multidimensional scaling. However, the assumptions made and the numerical methods used are not the same.

Correspondence analysis starts with data on the abundance of each of several characteristics for each of a set of objects. This is useful in ecology, for example, where the objects of interest are often different sites, the characteristics are different species, and the data consist of abundances of the species in samples taken from the sites. The purpose of correspondence analysis would then be to clarify the relationships between the sites as expressed by species distributions, and the relationships between the species as expressed by site distributions.

1.3   *The multivariate normal distribution*

The normal distribution for a single variable should be familiar to readers of this book. It has the well-known bell-shaped frequency curve, and many standard univariate statistical methods are based on the assumption that data are normally distributed.

Knowing the prominence of the normal distribution with univariate statistical methods, it will come as no surprise to discover that the multivariate normal distribution has a central position with multivariate statistical methods. Many of these methods require the assumption that the data being analyzed have multivariate normal distributions.

The exact definition of a multivariate normal distribution is not too important. The approach of most people, for better or worse, seems to be to regard data as being normally distributed unless there is some reason to believe that this is not true. In particular, if all the individual variables being studied appear to be normally distributed, then it is assumed that the joint distribution is multivariate normal. This is, in fact, a minimum requirement because the definition of multivariate normality requires more than this.

Cases do arise where the assumption of multivariate normality is clearly invalid. For example, one or more of the variables being studied may have a highly skewed distribution with several very high (or low) values; there may be many repeated values; etc. This type of problem can sometimes be overcome by an appropriate transformation of the data, as discussed in elementary texts on statistics. If this does not work, then a rather special form of analysis may be required.

One important aspect of a multivariate normal distribution is that it is specified completely by a mean vector and a covariance matrix. The definitions of a mean vector and a covariance matrix are given in Section 2.7. Basically, the mean vector contains the mean values for all of the variables being considered, while the covariance matrix contains the variances for all of the variables plus the covariances, which measure the extent to which all pairs of variables are related.

1.4 Computer programs

Practical methods for carrying out the calculations for multivariate analyses have been developed over about the last 70 years. However, the application of these methods for more than a small number of variables had to wait until computers became available. Therefore, it is only in the last 30 years or so that the methods have become reasonably easy to carry out for the average researcher.

Nowadays there are many standard statistical packages and computer programs available for calculations on computers of all types. It is intended that this book should provide readers with enough information to use any of these packages and programs intelligently, without saying much about any particular one. However, where it is appropriate, the software used to analyze example data will be mentioned.

1.5 Graphical methods

One of the outcomes of the greatly improved computer facilities in recent times has been an increase in the variety of graphical methods that are available for multivariate data. This includes contour plots and three-dimensional surface plots for functions of two variables, and a variety of special methods for showing the values that individual cases have for three or more variables. These methods are being used more commonly as part of the analysis of multivariate data, and they are therefore discussed at some length in Chapter 3.

1.6   Chapter summary

- Five data sets are introduced, and these will be used for examples throughout the remainder of the book. These data sets concern (1) five body measurements on female sparrows that did or did not survive a severe storm; (2) four measurements on skulls of Egyptian males living at five different periods in the past; (3) four measurements describing the environment and six measurements describing the genetic characteristics of 16 colonies of a butterfly in California and Oregon; (4) average values for six mandible measurements for seven canine groups, including prehistoric dogs from Thailand; and (5) percentages of people employed in nine different industry groups for 30 countries in Europe.
- Several important multivariate methods are briefly described in relationship to how they might be used with the data sets. These methods are principal components analysis, factor analysis, discriminant function analysis, cluster analysis, canonical correlation, multidimensional scaling, principal coordinates analysis, and correspondence analysis.
- The importance of the multivariate normal distribution is mentioned.
- The use of statistical packages is discussed, and it is noted that the individual packages used for example analyses will be mentioned where this is appropriate.
- The importance of graphical methods is noted.

References

Bumpus, H.C. (1898), The elimination of the unfit as illustrated by the introduced sparrow, *Passer domesticus, Biological Lectures*, 11th Lecture, Marine Biology Laboratory, Woods Hole, MA, pp. 209–226.

Euromonitor (1995), *European Marketing Data and Statistics*, Euromonitor Publications, London.

Higham, C.F.W., Kijngam, A., and Manly, B.F.J. (1980), An analysis of prehistoric canid remains from Thailand, *J. Archaeological Sci.*, 7, 149–165.

McKechnie, S.W., Ehrlich, P.R., and White, R.R. (1975), Population genetics of *Euphydryas* butterflies, I: genetic variation and the neutrality hypothesis, *Genetics*, 81: 571–594.

Thomson, A. and Randall-Maciver, R. (1905), *Ancient Races of the Thebaid*, Oxford University Press, Oxford, U.K.

United Nations (2000), *Statistical Yearbook*, 44th Issue, U.N. Department of Social Affairs, New York.

chapter two

Matrix algebra

2.1 *The need for matrix algebra*

The theory of multivariate statistical methods can be explained reasonably well only with the use of some matrix algebra. For this reason it is helpful, if not essential, to have at least some knowledge of this area of mathematics. This is true even for those who are interested in using the methods only as tools. At first sight, the notation of matrix algebra is somewhat daunting. However, it is not difficult to understand the basic principles, providing that some of the details are accepted on faith.

2.2 *Matrices and vectors*

An m × n *matrix* is an array of numbers with m rows and n columns, considered as a single entity, of the form:

$$\mathbf{A} = \begin{bmatrix} a_{11} & a_{12} & .. & a_{1n} \\ a_{21} & a_{22} & .. & a_{2n} \\ . & . & & . \\ . & . & & . \\ a_{m1} & a_{m2} & .. & a_{mn} \end{bmatrix}$$

If m = n, then it is a *square* matrix. If there is only one column, such as

$$\mathbf{c} = \begin{bmatrix} c_1 \\ c_2 \\ . \\ . \\ c_m \end{bmatrix}$$

then this is called a *column vector*. If there is only one row, such as

$$\mathbf{r} = (r_1, r_2, ..., r_n)$$

then this is called a *row vector*. Bold type is used to indicate matrices and vectors.

The *transpose* of a matrix is obtained by interchanging the rows and the columns. Thus the transpose of the matrix **A** above is

$$\mathbf{A'} = \begin{bmatrix} a_{11} & a_{21} & .. & a_{m1} \\ a_{12} & a_{22} & .. & a_{m2} \\ . & . & & . \\ . & . & & . \\ a_{1n} & a_{2n} & .. & a_{mn} \end{bmatrix}$$

Also, the transpose of the vector **c** is $\mathbf{c'} = (c_1, c_2, ..., c_m)$, and the transpose of the row vector **r** is the column vector **r'**.

There are a number of special kinds of matrices that are important. A *zero matrix* has all elements equal to zero, so that it is of the form

$$\mathbf{0} = \begin{bmatrix} 0 & 0 & .. & 0 \\ 0 & 0 & .. & 0 \\ . & . & & . \\ . & . & & . \\ 0 & 0 & .. & 0 \end{bmatrix}$$

A *diagonal matrix* has zero elements except down the main diagonal, so that it takes the form

$$\mathbf{D} = \begin{bmatrix} d_1 & 0 & .. & 0 \\ 0 & d_2 & .. & 0 \\ . & . & & . \\ . & . & & . \\ 0 & 0 & .. & d_n \end{bmatrix}$$

A *symmetric matrix* is a square matrix that is unchanged when it is transposed, so that $\mathbf{A'} = \mathbf{A}$. Finally, an *identity matrix* is a diagonal matrix with all on the diagonal terms equal to one, so that

$$I = \begin{bmatrix} 1 & 0 & .. & 0 \\ 0 & 1 & .. & 0 \\ . & . & & . \\ . & . & & . \\ 0 & 0 & .. & 1 \end{bmatrix}$$

Two matrices are *equal* only if they are the same size and all of their elements are equal. For example

$$\begin{bmatrix} a_{11} & a_{12} & a_{13} \\ a_{21} & a_{22} & a_{23} \\ a_{31} & a_{32} & a_{33} \end{bmatrix} = \begin{bmatrix} b_{11} & b_{12} & b_{13} \\ b_{21} & b_{22} & b_{23} \\ b_{31} & b_{32} & b_{33} \end{bmatrix}$$

only if $a_{11} = b_{11}$, $a_{12} = b_{12}$, $a_{13} = b_{13}$, and so on.

The *trace* of a matrix is the sum of the diagonal terms, which is only defined for a square matrix. For example, the trace of the 3×3 matrix with the elements a_{ij} shown above has trace $(\mathbf{A}) = a_{11} + a_{22} + a_{33}$.

2.3   *Operations on matrices*

The ordinary arithmetic processes of addition, subtraction, multiplication, and division have their counterparts with matrices. With addition and subtraction, it is just a matter of working element by element with two matrices of the same size. For example, if A and B are both of size 3×2, then

$$\mathbf{A} + \mathbf{B} = \begin{bmatrix} a_{11} & a_{12} \\ a_{21} & a_{22} \\ a_{31} & a_{32} \end{bmatrix} + \begin{bmatrix} b_{11} & b_{12} \\ b_{21} & b_{22} \\ b_{31} & b_{32} \end{bmatrix} = \begin{bmatrix} a_{11}+b_{11} & a_{12}+b_{12} \\ a_{21}+b_{21} & a_{22}+b_{22} \\ a_{31}+b_{31} & a_{32}+b_{32} \end{bmatrix}$$

while

$$\mathbf{A} - \mathbf{B} = \begin{bmatrix} a_{11} & a_{12} \\ a_{21} & a_{22} \\ a_{31} & a_{32} \end{bmatrix} - \begin{bmatrix} b_{11} & b_{12} \\ b_{21} & b_{22} \\ b_{31} & b_{32} \end{bmatrix} = \begin{bmatrix} a_{11}-b_{11} & a_{12}-b_{12} \\ a_{21}-b_{21} & a_{22}-b_{22} \\ a_{31}-b_{31} & a_{32}-b_{32} \end{bmatrix}$$

Clearly, these operations only apply with two matrices of the same size.

In matrix algebra, an ordinary number such as 20 is called a *scalar*. Multiplication of a matrix **A** by a scalar k is then defined to be the multiplication of every element in **A** by k. Thus if **A** is the 3×2 matrix as shown above, then

$$kA = \begin{bmatrix} ka_{11} & ka_{12} \\ ka_{21} & ka_{22} \\ ka_{31} & ka_{32} \end{bmatrix}$$

The multiplication of two matrices, denoted by **A.B** or **A** × **B**, is more complicated. To begin with, **A.B** is defined only if the number of columns of **A** is equal to the number of rows of **B**. Assume that this is the case, with **A** having the size m × n and **B** having the size n × p. Then multiplication is defined to produce the result:

$$\mathbf{A.B} = \begin{bmatrix} a_{11} & a_{12} & .. & a_{1n} \\ a_{21} & a_{22} & .. & a_{2n} \\ . & . & & . \\ . & . & & . \\ a_{m1} & a_{m2} & .. & a_{mn} \end{bmatrix} . \begin{bmatrix} b_{11} & b_{12} & .. & b_{1p} \\ b_{21} & b_{22} & .. & b_{2p} \\ . & . & & . \\ . & . & & . \\ b_{n1} & b_{n2} & .. & b_{np} \end{bmatrix}$$

$$= \begin{bmatrix} \Sigma a_{1j}b_{j1} & \Sigma a_{1j}b_{j2} & .. & \Sigma a_{1j}b_{jp} \\ \Sigma a_{2j}b_{j1} & \Sigma a_{2j}b_{j2} & .. & \Sigma a_{2j}b_{jp} \\ . & . & & . \\ . & . & & . \\ \Sigma a_{mj}b_{j1} & \Sigma a_{mj}b_{j2} & .. & \Sigma a_{mj}b_{jp} \end{bmatrix}$$

where the summations are for j running from 1 to n. Hence the element in the ith row and kth column of **A.B** is

$$\Sigma \, a_{ij} \, b_{jk} = a_{i1} \, b_{1k} + a_{i2} \, b_{2k} + .. + a_{in} \, b_{nk}$$

When A and B are both square matrices, then **A.B** and **B.A** are both defined. However, they are not generally equal. For example,

$$\begin{bmatrix} 2 & -1 \\ 1 & 1 \end{bmatrix} . \begin{bmatrix} 1 & 1 \\ 0 & 1 \end{bmatrix} = \begin{bmatrix} 2\times1-1\times0 & 2\times1-1\times1 \\ 1\times1+1\times0 & 1\times1+1\times1 \end{bmatrix} = \begin{bmatrix} 2 & 1 \\ 1 & 2 \end{bmatrix}$$

whereas

$$\begin{bmatrix} 1 & 1 \\ 0 & 1 \end{bmatrix} \cdot \begin{bmatrix} 2 & -1 \\ 1 & 1 \end{bmatrix} = \begin{bmatrix} 1\times2+1\times1 & -1\times1+1\times1 \\ 0\times2+1\times1 & -1\times0+1\times1 \end{bmatrix} = \begin{bmatrix} 3 & 0 \\ 1 & 1 \end{bmatrix}$$

2.4 Matrix inversion

Matrix inversion is analogous to the ordinary arithmetic process of division. For a scalar k, it is of course true that $k \times k^{-1} = 1$. In a similar way, if A is a square matrix and

$$A \times A^{-1} = I$$

where I is the identity matrix, then the matrix A^{-1} is the *inverse* of the matrix A. Inverses exist only for square matrices, but all square matrices do not have inverses. If an inverse does exist, then it is both a left inverse, so that $A^{-1} \times A = I$, as well as a right inverse so that $A \times A^{-1} = I$.

An example of an inverse matrix is

$$\begin{bmatrix} 2 & 1 \\ 1 & 2 \end{bmatrix}^{-1} = \begin{bmatrix} 2/3 & -1/3 \\ -1/3 & 2/3 \end{bmatrix}$$

which can be verified by checking that

$$\begin{bmatrix} 2 & 1 \\ 1 & 2 \end{bmatrix} \cdot \begin{bmatrix} 2/3 & -1/3 \\ -1/3 & 2/3 \end{bmatrix} = \begin{bmatrix} 1 & 0 \\ 0 & 1 \end{bmatrix}$$

Actually, the inverse of a 2×2 matrix, if it exists, can be calculated fairly easily. The equation is

$$\begin{bmatrix} a & b \\ c & d \end{bmatrix}^{-1} = \begin{bmatrix} d/\Delta & -b/\Delta \\ -c/\Delta & a/\Delta \end{bmatrix}$$

where $\Delta = (a \times d) - (b \times c)$. Here the scalar Δ is called the *determinant* of the matrix being inverted. Clearly, the inverse is not defined if $\Delta = 0$ because finding the elements of the inverse then involves a division by zero. For 3×3 and larger matrices, the calculation of the inverse is tedious and is best done by using a computer program. Nowadays even spreadsheets include a facility to compute an inverse.

Any square matrix has a determinant, which can be calculated by a generalization of the equation just given for the 2×2 case. If the determinant is zero, then the inverse does not exist, and vice versa. A matrix with a zero determinant is said to be *singular*.

Matrices sometimes arise for which the inverse is equal to the transpose. They are then said to be *orthogonal*. Hence **A** is orthogonal if $\mathbf{A}^{-1} = \mathbf{A}'$.

2.5   Quadratic forms

Suppose that **A** is an $n \times n$ matrix and **x** is a column vector of length n. Then the quantity

$$Q = \mathbf{x}' \, \mathbf{A} \, \mathbf{x}$$

is a scalar that is called a *quadratic form*. This scalar can also be expressed as

$$Q = \sum_{i=1}^{n} \sum_{j=1}^{n} x_i a_{ij} x_j$$

where x_i is the element in the ith row of **x** and a_{ij} is the element in the ith row and jth column of **A**.

2.6   Eigenvalues and eigenvectors

Consider the set of linear equations

$$a_{11}x_1 + a_{12}x_2 + \ldots + a_{1n}x_n = \lambda x_1$$
$$a_{21}x_1 + a_{22}x_2 + \ldots + a_{2n}x_n = \lambda x_2$$
$$\cdot$$
$$a_{n1}x_1 + a_{n2}x_2 + \ldots + a_{nn}x_n = \lambda x_n$$

where λ is a scalar. These can also be written in matrix form as

$$\mathbf{A} \, \mathbf{x} = \lambda \mathbf{x}$$

or

$$(\mathbf{A} - \lambda \, \mathbf{I}) \, \mathbf{x} = \mathbf{0}$$

where **I** is the $n \times n$ identity matrix, and **0** is an $n \times 1$ vector of zeros. Then it can be shown that these equations can hold only for certain particular values of λ that are called the *latent roots* or *eigenvalues* of **A**. There can be up to n of these eigenvalues. Given the ith eigenvalue λ_i, the equations can be solved by arbitrarily setting $x_1 = 1$, and the resulting vector of x values with transpose $x' = (1, x_2, x_3, .., x_n)$, or any multiple of this vector, is called the ith *latent root* or the ith *eigenvector* of the matrix **A**. Also, the sum of the eigenvalues is equal to the trace of **A** defined above, so that

$$\text{trace } (\mathbf{A}) = \lambda_1 + \lambda_2 + .. + \lambda_n$$

2.7   *Vectors of means and covariance matrices*

Population and sample values for a single random variable are often summarized by the values for the mean and variance. Thus if a sample of size n yields the values $x_1, x_2, .., x_n$, then the *sample mean* is defined to be

$$\bar{x} = \left(x_1 + x_2 + ... + x_n\right)/n = \sum_{i=1}^{n} x_i/n$$

while the *sample variance* is

$$s^2 = \sum_{i=1}^{n} \left(x_i - \bar{x}\right)^2 /(n-1)$$

These are estimates of the corresponding population parameters, which are the *population mean* μ and the *population variance* σ^2.

In a similar way, multivariate populations and samples can be summarized by *mean vectors* and *covariance matrices*. Suppose that there are p variables $X_1, X_2, .., X_p$ being considered, and that a sample of n values for each of these variables is available. Let the sample mean and sample variance for the ith variable be $\bar{x}_i$ and s_i^2, respectively, where these are calculated using the equations given above. In addition, the *sample covariance* between variables X_j and X_k is

$$c_{jk} = \sum_{i=1}^{n} \left(x_{ij} - \bar{x}_j\right)\left(x_{jk} - \bar{x}_k\right)/(n-1)$$

where x_{ij} is the value of variable X_j for the ith multivariate observation. This covariance is then a measure of the extent to which there is a linear relationship between X_j and X_k, with a positive value indicating that large value of X_j and X_k tend to occur together, and a negative value indicating that large values for one variable tend to occur with small values for the other variable. It is related to the ordinary correlation coefficient between the two variables, which is defined to be

$$r_{jk} = c_{jk}/(s_j\, s_k)$$

Furthermore, the definitions imply that $c_{kj} = c_{jk}$, $r_{kj} = r_{jk}$, $c_{jj} = s_j^2$, and $r_{jj} = 1$. With these definitions, the transpose of the *sample mean vector* is

$$\bar{x}' = (\bar{x}_1, \bar{x}_2, .., \bar{x}_p)$$

which can be thought of as reflecting the center of the multivariate sample. It is also an estimate of the transpose of the population vector of means

$$\mu' = (\mu_1, \mu_2, .., \mu_p)$$

Furthermore, the sample matrix of variances and covariances, or the *covariance matrix*, is

$$C = \begin{bmatrix} c_{11} & c_{12} & .. & c_{1p} \\ c_{21} & c_{22} & .. & c_{2p} \\ . & . & & . \\ . & . & & . \\ c_{p1} & c_{p2} & .. & c_{pp} \end{bmatrix}$$

where $c_{ii} = s_i^2$. This is also sometimes called the *sample dispersion matrix*, and it measures the amount of variation in the sample as well as the extent to which the p variables are correlated. It is an estimate of the *population covariance matrix*

$$\Sigma = \begin{bmatrix} \sigma_{11} & \sigma_{12} & .. & \sigma_{1p} \\ \sigma_{21} & \sigma_{22} & .. & \sigma_{2p} \\ . & . & & . \\ . & . & & . \\ \sigma_{p1} & \sigma_{p2} & .. & \sigma_{pp} \end{bmatrix}$$

Finally, the sample correlation matrix is

$$R = \begin{bmatrix} 1 & r_{12} & .. & r_{1p} \\ r_{21} & 1 & .. & r_{2p} \\ . & . & & . \\ . & . & & . \\ r_{p1} & r_{p2} & .. & 1 \end{bmatrix}$$

Again, this is an estimate of the corresponding *population correlation matrix*. An important result for some analyses is that if the observations for each of the variables are coded by subtracting the sample mean and dividing by the sample standard deviation, then the coded values will have a mean of zero and a standard deviation of one for each variable. In that case, the sample covariance matrix will equal the sample correlation matrix, i.e., $\mathbf{C} = \mathbf{R}$.

2.8   Further reading

This short introduction to matrix algebra will suffice for understanding the methods described in the remainder of this book and some of the theory behind these methods. However, for a better understanding of the theory, more knowledge and proficiency is required.

One possibility in this respect is just to read a university text giving an introduction to matrix methods. Alternatively, there are several books of various lengths that cover what is needed just for statistical applications. Three of these are by Searle (1982), Healy (1986), and Harville (1997), with Healy's book being quite short (less than 100 pages), Searle's book quite long (438 pages), and Harville's book the longest of all (630 pages). Another short book with less than 100 pages is by Namboodiri (1984). The shorter books should be more than adequate for most people.

Another possibility is to do a Web search on the topic of matrix algebra. This yields much educational material including free books and course notes.

2.9   Chapter summary

- Matrices and vectors are defined, as are the special forms of the zero, diagonal, and identity matrices. Definitions are also given for the transpose of a matrix, the equality of two matrices, and the trace of a matrix.
- The operations of addition, subtraction, and multiplication are defined for two matrices.
- The meaning of matrix inversion is briefly explained, together with the associated concepts of a determinant, a singular matrix, and an orthogonal matrix.
- A quadratic form is defined.
- Eigenvalues and eigenvectors (latent roots and vectors) are defined.
- The calculation of the sample mean vector and the sample covariance matrix are explained, together with the corresponding population mean vector and population covariance matrix. The sample correlation matrix and the corresponding population correlation matrix are also defined.
- Suggestions are made about books and other sources of further information about matrix algebra.

References

Harville, D.A. (1997), *Matrix Algebra from a Statistician's Perspective*, Springer, New York.

Healy, M.J.R. (1986), *Matrices for Statistics*, Clarendon Press, Oxford.

Namboodiri, K. (1984), *Matrix Algebra: an Introduction*, Sage Publications, Thousand Oaks, CA.

Searle, S.R. (1982), *Matrix Algebra Useful to Statisticians*, Wiley, New York.

chapter three

Displaying multivariate data

3.1 The problem of displaying many variables in two dimensions

Graphs must be displayed in two dimensions either on paper or on a computer screen. It is therefore straightforward to show one variable plotted on a vertical axis against a second variable plotted on a horizontal axis. For example, Figure 3.1 shows the alar extent plotted against the total length for the 49 female sparrows measured by Hermon Bumpus in his study of natural selection (Table 1.1). Such plots allow one or more other characteristics of the objects being studied to be shown as well. For example, in the case of Bumpus's sparrows, survival and nonsurvival are also indicated.

It is considerably more complicated to show one variable plotted against another two, but still possible. Thus Figure 3.2 shows beak and head lengths (as a single variable) plotted against total lengths and alar lengths for the 49 sparrows. Again, different symbols are used for survivors and nonsurvivors.

It is not possible to show one variable plotted against another three at the same time in some extension of a three-dimensional plot. Hence there is a major problem in showing in a simple way the relationships that exist between the individual objects in a multivariate set of data where those objects are each described by four or more variables. Various solutions to this problem have been proposed and are discussed in this chapter.

3.2 Plotting index variables

One approach to making a graphical summary of the differences between objects that are described by more than four variables involves plotting the objects against the values of two or three index variables. Indeed, a major objective of many multivariate analyses is to produce index variables that can be used for this purpose, a process that is sometimes called ordination. For example, a plot of the values of principal component 2 against the values

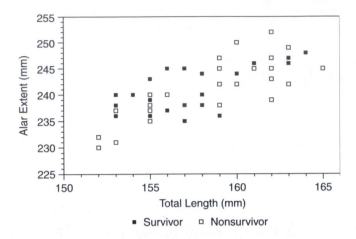

Figure 3.1 Alar extent plotted against total length for the 49 female sparrows measured by Hermon Bumpus.

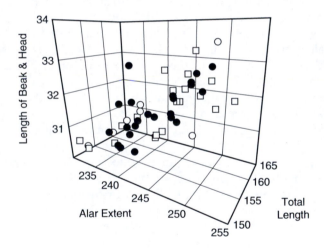

Figure 3.2 The length of the beak and head plotted against the total length and alar extent (all measured in millimeters) for the 49 female sparrows measured by Hermon Bumpus (• = survivor, o = nonsurvivor).

of principal component 1 can be used as a means of representing the relationships between objects graphically, and a display of principal component 3 against the first two principal components can also be used if necessary.

The use of suitable index variables has the advantage of reducing the problem of plotting many variables to two or three dimensions, but the potential disadvantage is that some key difference between the objects may be lost in the reduction. This approach is discussed in various different contexts in the chapters that follow and will not be considered further here.

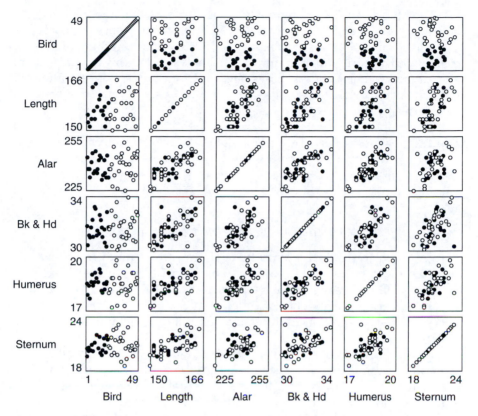

Figure 3.3 Draftsman's plot of the bird number and five variables measured (in millimeters) on 49 female sparrows. The variables are the total length, the alar extent, the length of the beak and head, the length of the humerus, and the length of the keel of the sternum (• = survivor, o = nonsurvivor). Only the extreme values are shown on each scale.

3.3 The draftsman's plot

A draftsman's display of multivariate data consists of a plot of the values for each variable against the values for each of the other variables, with the individual graphs being small enough so that they can all be viewed at the same time. This has the advantage of only needing two-dimensional plots, but the disadvantage is that it cannot depict aspects of the data that would only be apparent when three or more variables are considered together.

An example is shown in Figure 3.3. Here, the five variables measured by Hermon Bumpus on 49 sparrows (total length, alar extent, length of beak and head, length of humerus, and length of the keel of the sternum, all in mm) are plotted for the data given in Table 1.1, with an additional first variable being the number of the sparrow, from 1 to 49. Different symbols are used for the measurements on survivors (birds 1 to 21) and nonsurvivors (birds 22 to 49). Regression lines are also sometimes added to the plots.

This type of plot is obviously good for showing the relationships between pairs of variables and for highlighting the existence of any objects that have unusual values for one or two variables. It can therefore be recommended as part of many multivariate analyses, and this feature is available in many statistical packages, sometimes as what is called a scatter-plot matrix. Some packages also allow the option of specifying the horizontal and vertical variables without insisting that these be the same.

The individual objects are not easily identified on a draftsman's plot, and it is therefore usually not immediately clear which objects are similar and which are different. Therefore, this type of plot is not suitable for showing relationships between objects, as distinct from relationships between variables.

3.4   The representation of individual data points

An approach to displaying data that is more truly multivariate involves representing each of the objects for which variables are measured by a symbol, with different characteristics of this symbol varying according to different variables. A number of different types of symbol have been proposed for this purpose including faces (Chernoff, 1973) and stars (Welsch, 1976).

As an illustration, consider the data in Table 1.4 on mean values of six mandible measurements for seven canine groups, as discussed in Example 1.4. Here an important question concerns which of the other groups is most similar to the prehistoric Thai dog, and it can be hoped that this becomes apparent from a graphical comparison of the groups. To this end, Figure 3.4 shows the data represented by faces and stars.

For the faces, there was the following connection between features and the variables: mandible breadth to eye size, mandible height to nose size, length of first molar to brow size, breadth of first molar to ear size, length from first to third molar to mouth size, and length from first to fourth premolars to the amount of smile. For example, the eyes are largest for the Chinese wolf with the maximum mandible breadth of 13.5 mm, and smallest for the golden jackal with the minimum mandible breadth of 8.1 mm. It is apparent from the plots that prehistoric Thai dogs are most similar to modern Thai dogs, and most different from Chinese wolves.

For the stars, the six variables were assigned to rays in the order (1) mandible breadth, (2) mandible height, (3) length of first molar, (4) breadth of first molar, (5), length from first to third molar, and (6) length from first to fourth premolars. The mandible length is represented by the ray corresponding to six o'clock and the other variables follow in a clockwise order as indicated by the key that accompanies the figure. Inspection of the stars indicates again that the prehistoric Thai dogs are most similar to modern Thai dogs and most different from Chinese wolves.

Suggestions for alternatives to faces and stars, and a discussion of the relative merits of different symbols, are provided by Everitt (1978) and

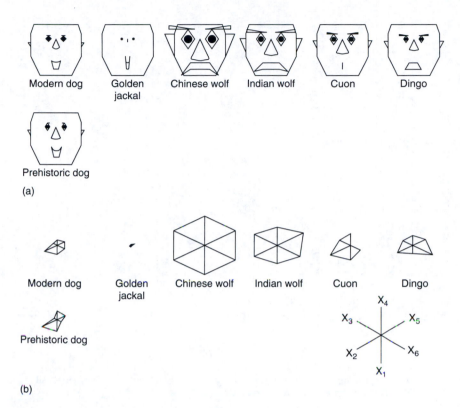

Figure 3.4 Graphical representation of mandible measurements on different canine groups using (a) Chernoff faces and (b) stars. (For definitions of variables X_1 through X_6, see caption for Figure 3.5.)

Toit et al. (1986, ch. 4). In summary, it can be said that the use of symbols has the advantage of displaying all variables simultaneously, but the disadvantage that the impression gained from the graph may depend quite strongly on the order in which objects are displayed and the order in which variables are assigned to the different aspects of the symbol.

The assignment of variables is likely to have more effect with faces than it is with stars, because variation in different features of the face may have very different impacts on the observer, whereas this is less likely to be the case with different rays of a star. For this reason, the recommendation is often made that alternative assignments of variables to features should be tried with faces in order to find what seems to be the best. The subjective nature of this type of process is clearly rather unsatisfactory.

Although the use of faces, stars, and other similar representations for the values of variables on the objects being considered seems to be useful under some circumstances, the fact is that this is seldom done. One reason is the difficulty in finding computer software to produce the graphs. In the past, this software was reasonably easily available, but these options are now very hard to find in statistical packages.

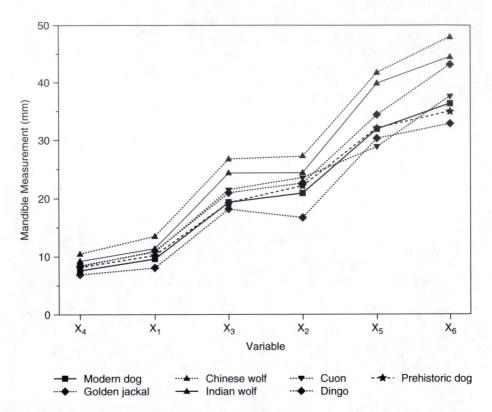

Figure 3.5 Profiles of variables for mandible measurements on seven canine groups The variables are in order of increasing average values, with X_1 = breadth of mandible; X_2 = height of mandible above the first molar; X_3 = length of first molar; X_4 = breadth of first molar; X_5 = length from first to third molar, inclusive; and X_6 = length from first to fourth molar, inclusive.

3.5 *Profiles of variables*

Another way to represent objects that are described by several measured variables is through lines that show the profile of variable values. A simple way to draw these involves just plotting the values for the variables, as shown in Figure 3.5 for the seven canine groups that have already been considered. The similarity between prehistoric and modern Thai dogs noted from the earlier graphs is still apparent, as is the difference between prehistoric dogs and Chinese wolves. In this, graph the variables have been plotted in the order of their average values for the seven groups to help in emphasizing similarities and differences.

An alternative representation using bars instead of lines is shown in Figure 3.6. Here the variables are in their original order because there seems little need to change this when bars are used. The conclusion about similarities

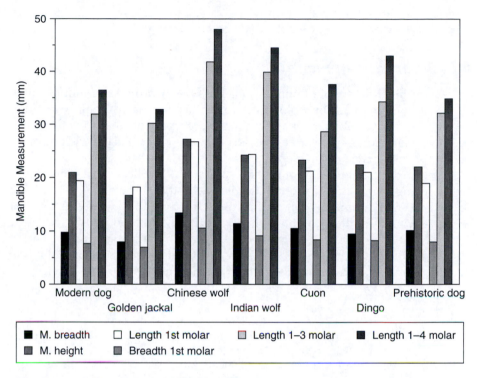

Figure 3.6 An alternative way to show variable profiles using bars instead of the lines used in Figure 3.5 (M. = mandible).

and differences between the canine groups is exactly the same as seen in Figure 3.5.

3.6 Discussion and further reading

It seems fair to say that there is no method for displaying data on many variables at a time that is completely satisfactory in situations when it is not desirable to reduce these variables to two or three index variables (using one of the methods to be discussed later in this book). The three types of method that have been discussed here involve the use of a draftsman's display with all pairs of variables plotted against each other, symbols (stars or faces), and profiles of variables. Which of these is most suitable for a particular application depends on the circumstances, but as a general rule the draftsman's display is good for highlighting relationships between pairs of variables, while the use of symbols or profiles is good for highlighting unusual cases and similar cases.

For further information about the theory of the construction of graphs in general, see the books by Cleveland (1985) and Tufte (2001). More details on graphical methods specifically for multivariate data are described in books by Everitt (1978), Toit et al. (1986), and Jacoby (1998).

3.7   Chapter summary

- The difficulty of plotting results for several variables simultaneously is discussed. One solution is to replace several variables with one index variable, such as a principal component, that in some sense summarizes the variation in all of the variables. This process is sometimes called ordination.
- A draftsman's plot for several variables is a collection of plots showing each variable plotted against each other variable.
- When individual data points are defined by their values for several variables, there are many methods that have proposed to represent these individual points to visualize which are similar and which are different in terms of the variables. Chernoff faces, star plots, line profiles, and bar profiles are illustrated for this purpose.

References

Chernoff, H. (1973), Using faces to represent points in K-dimensional space graphically, *J. Am. Stat. Assoc.*, 68, 361–368.

Cleveland, W.S. (1985), *The Elements of Graphing Data*, Wadsworth, Monterey, CA.

Everitt, B. (1978), *Graphical Techniques for Multivariate Data*, North-Holland, New York.

Jacoby, W.G. (1998), *Statistical Graphics for Visualizing Multivariate Data*, Sage Publications, Thousand Oaks, CA.

Toit, S.H.C., Steyn, A.G.W., and Stumf, R.H. (1986), *Graphical Exploratory Data Analysis*, Springer-Verlag, New York.

Tufte, E.R. (2001), *The Visual Display of Quantitative Information*, 2nd ed., Graphics Press, Cheshire, CT.

Welsch, R.E. (1976), Graphics for data analysis, *Comput. Graphics*, 2, 31–37.

chapter four

Tests of significance with multivariate data

4.1 Simultaneous tests on several variables

When data are collected for several variables on the same sample units, then it is always possible to examine the variables one at a time as far as tests of significance are concerned. For example, if the sample units are in two groups, then a difference between the means for the two groups can be tested separately for each variable. Unfortunately, there is a drawback to this simple approach in that it requires the repeated use of significance tests, each of which has a certain probability of leading to a wrong conclusion. As will be discussed further in Section 4.4, the probability of falsely finding at least one significant difference accumulates with the number of tests carried out, so that it may become unacceptably large.

There are ways of adjusting significance levels to allow for many tests being carried out at the same time, but it may be preferable to conduct a single test that uses the information from all variables together. For example, it might be desirable to test the hypothesis that the means of all variables are the same for two multivariate populations, with a significant result being taken as evidence that the means differ for at least one variable. These types of overall test are considered in this chapter for the comparison of means and variation for two or more samples.

4.2 Comparison of mean values for two samples: the single variable case

Consider the data in Table 1.1 on the body measurements of 49 female sparrows. Consider in particular the first measurement, which is total length. A question of some interest might be whether the mean of this variable was the same for survivors and nonsurvivors of the storm that led to the birds' being collected. There is then a sample of 21 survivors and a second sample of 28 nonsurvivors. Assuming that these are effectively random samples from much larger populations of survivors and nonsurvivors, the question then is whether the two sample means are significantly different in the sense that the observed mean difference is so large that it is unlikely to have occurred if the population means are equal. A standard approach would be to carry out a t-test.

Thus, suppose that in a general situation there is a single variable X and two random samples of values are available from different populations. Let x_{i1} denote the values of X in the first sample, for $i = 1, 2, ..., n_1$, and x_{i2} denote the values in the second sample, for $i = 1, 2, ..., n_2$. Then the mean and variance for the jth sample are, respectively,

$$\bar{x}_j = \sum_{i=1}^{n_j} x_{ij}/n_j$$

and

$$s_j^2 = \sum_{i=1}^{n_j} \left(x_{ij} - \bar{x}_j\right)^2 \Big/ \left(n_j - 1\right) \tag{4.1}$$

On the assumption that X is normally distributed in both samples, with a common within-sample variance, a test to see whether the two sample means are significantly different involves calculating the statistic

$$t = \left(\bar{x}_1 - \bar{x}_2\right) \left\{s\sqrt{\left(1/n_1 + 1/n_2\right)}\right\} \tag{4.2}$$

and seeing whether this is significantly different from zero in comparison with the t-distribution with $n_1 + n_2 - 2$ degrees of freedom (df). Here

$$s^2 = \left\{\left(n_1 - 1\right)s_1^2 + \left(n_2 - 1\right)s_2^2\right\} \Big/ \left(n_1 + n_2 - 2\right) \tag{4.3}$$

is the pooled estimate of variance from the two samples.

It is known that this test is fairly robust to the assumption of normality, so if the population distributions of X are not too different from normal, it should be satisfactory, particularly for sample sizes of about 20 or more. The assumption of equal population variances is also not too crucial if the ratio of the true variances is within the limits 0.4 to 2.5. The test is particularly robust if the two sample sizes are equal, or nearly so (Carter et al., 1979).

If there are no concerns about nonnormality but the population variances may be quite unequal, then one possibility is to use a modified t-test. For example, Welch's (1951) test can be used, where this has the test statistic

$$t = \left(\bar{x}_1 - \bar{x}_2\right) \Big/ \left\{\sqrt{\left(s_1^2/n_1 + s_2^2/n_2\right)}\right\} \tag{4.4}$$

Evidence for unequal population means is then obtained if t is significantly different from zero in comparison with the t-distribution with df equal to

$$v = (w_1 + w_2)^2 / \{w_1^2/(n_1 - 1) + w_2^2/(n_2 - 1)\} \tag{4.5}$$

where $w_1 = s_1^2/n_1$ and $w_2 = s_2^2/n_2$.

When there are concerns about both nonnormality and unequal variances, then it has been shown by Manly and Francis (2002) that it may not be possible to reliably test for a difference in the population means. In particular, there may be too many significant results providing evidence for a population mean difference when this does not really exist, irrespective of what test procedure is used. Manly and Francis provided a solution for this problem via a testing scheme that includes (a) an assessment of whether two or more samples differ with respect to means or variances using robust randomization tests (Manly, 2001, sec. 4.6) and (b) an assessment of whether the randomization test for mean differences is reliable. See their paper for more details.

4.3   Comparison of mean values for two samples: the multivariate case

Consider again the female sparrow data that are shown in Table 1.1. The t-test described in the previous section can obviously be employed for each of the five measurements shown in the table (total length, alar extent, length of beak and head, length of humerus, and length of keel of sternum). In that way, it is possible to decide which, if any, of these variables appear to have had different mean values for the populations of survivors and nonsurvivors. However, in addition to these tests, it may also be of some interest to know whether all five variables considered together suggest a difference between survivors and nonsurvivors. In other words, does the total evidence point to mean differences between the populations of surviving and nonsurviving sparrows?

What is needed to answer this question is a multivariate test. One possibility is Hotelling's T^2-test. The statistic used is then a generalization of the t-statistic of Equation 4.2 or, to be more precise, the square of this t-statistic.

In a general case, there will be p variables $X_1, X_2, ..., X_p$ being considered and two samples with sizes n_1 and n_2. There are then two sample mean vectors, $\bar{x}_1$ and $\bar{x}_2$, and two sample covariance matrices, C_1 and C_2, with all of these being calculated as explained in Section 2.7.

Assuming that the population covariance matrices are the same for both populations, a pooled estimate of this matrix is

$$C = \{(n_1 - 1)C_1 = (n_2 - 1)C_2\}/(n_1 + n_2 - 2) \tag{4.6}$$

Hotelling's T²-statistic is then defined as

$$T^2 = n_1 n_2 (\bar{x}_1 - \bar{x}_2)' C^{-1} (\bar{x}_1 - \bar{x}_2)/(n_1 + n_2) \qquad (4.7)$$

A significantly large value for this statistic is evidence that the two population mean vectors are different. The significance or lack of significance of T² is most simply determined by using the fact that if the null hypothesis of equal population mean vectors is true, then the transformed statistic

$$F = (n_1 + n_2 - p - 1) T^2 / \{(n_1 + n_2 - 2)p\} \qquad (4.8)$$

follows an F-distribution with p and $(n_1 + n_2 - p - 1)$ df.

The T²-statistic is a quadratic form, as defined in Section 2.5. It can therefore be written as the double sum

$$T^2 = \{(n_1 n_2)/(n_1 + n_2)\} \sum_{i=1}^{p} \sum_{k=1}^{p} (\bar{x}_{1i} - \bar{x}_{2i}) c^{ik} (\bar{x}_{1k} - \bar{x}_{2k}) \qquad (4.9)$$

which may be simpler to compute. Here $\bar{x}_{ji}$ is the mean of variable X_i in the jth sample, and c^{ik} is the element in the ith row and kth column of the inverse matrix C^{-1}.

The two samples being compared using the T²-statistic are assumed to come from multivariate normal distributions with equal covariance matrices. Some deviation from multivariate normality is probably not serious. A moderate difference between population covariance matrices is also not too important, particularly with equal or nearly equal sample sizes (Carter et al., 1979). If the two population covariance matrices are very different, and if sample sizes are very different as well, then a modified test can be used (Yao, 1965), but this still relies on the assumption of multivariate normality.

Example 4.1   Testing mean values for Bumpus's female sparrows

As an example of the use of the univariate and multivariate tests that have been described for two samples, consider again the female sparrow data from Table 1.1. Here it is a question of whether there is any difference between survivors and nonsurvivors with respect to the mean values of five morphological characters.

First of all, tests on the individual variables can be considered, starting with X_1, the total length. The mean of this variable for the 21 survivors is $\bar{x}_1 = 157.38$, while the mean for the 28 nonsurvivors is $\bar{x}_2 = 158.43$. The corresponding sample variances are $s_1^2 = 11.05$ and $s_2^2 = 15.07$. The pooled variance from Equation 4.3 is therefore

$$s^2 = (20 \times 11.05 + 27 \times 15.07)/47 = 13.36$$

and the t-statistic of Equation 4.2 is

$$t = (157.38 - 158.43)/\sqrt{\{13.36(1/21 + 1/28)\}} = -0.99$$

with $n_1 + n_2 - 2 = 47$ df. This is not significantly different from zero at the 5% level, so there is no evidence of a population mean difference between survivors and nonsurvivors with regard to total length.

Table 4.1 summarizes the results of tests on all five of the variables taken individually. In no case is there any evidence of a population mean difference between survivors and nonsurvivors.

For tests on all five variables considered together it is necessary to know the sample mean vectors and covariance matrices. The means are given in Table 4.1, and the covariance matrices are defined in Section 2.7. For the sample of 21 survivors, the mean vector and covariance matrix are:

$$\bar{x}_1 = \begin{bmatrix} 157.381 \\ 241.000 \\ 31.433 \\ 18.500 \\ 20.810 \end{bmatrix} \text{ and } C_1 = \begin{bmatrix} 11.048 & 9.100 & 1.557 & 0.870 & 1.286 \\ 9.100 & 17.500 & 1.910 & 1.310 & 0.880 \\ 1.557 & 1.910 & 0.531 & 0.189 & 0.240 \\ 0.870 & 1.310 & 0.189 & 0.176 & 0.133 \\ 1.286 & 0.880 & 0.240 & 0.133 & 0.575 \end{bmatrix}$$

For the sample of 28 nonsurvivors, the results are

$$\bar{x}_2 = \begin{bmatrix} 158.429 \\ 241.571 \\ 31.479 \\ 18.446 \\ 20.839 \end{bmatrix} \text{ and } C_2 = \begin{bmatrix} 15.069 & 17.190 & 2.243 & 1.746 & 2.931 \\ 17.190 & 32.550 & 3.398 & 2.950 & 4.066 \\ 2.243 & 3.398 & 0.728 & 0.470 & 0.559 \\ 1.746 & 2.950 & 0.470 & 0.434 & 0.506 \\ 2.931 & 4.066 & 0.559 & 0.506 & 1.321 \end{bmatrix}$$

Table 4.1 Comparison of Mean Values for Survivors and Nonsurvivors for Bumpus's Female Sparrows with Variables Taken One at a Time

Variable	Survivors		Nonsurvivors		t (47 df)	P-value[a]
	$\bar{x}_1$	s_1^2	$\bar{x}_2$	s_2^2		
Total length	157.38	11.05	158.43	15.07	−0.99	0.327
Alar extent	241.00	17.50	241.57	32.55	−0.39	0.698
Length of beak and head	31.43	0.53	31.48	0.73	−0.20	0.842
Length of humerus	18.50	0.18	18.45	0.43	0.33	0.743
Length of keel of sternum	20.81	0.58	20.84	1.32	−0.10	0.921

[a] Probability of obtaining a t-value as far from zero as the observed value if the null hypothesis of no population mean difference is true.

The pooled sample covariance matrix is then

$$C = (20\ \mathbf{C_1} + 27\ \mathbf{C_2})/47 = \begin{bmatrix} 13.358 & 13.748 & 1.951 & 1.373 & 2.231 \\ 13.748 & 26.146 & 2.765 & 2.252 & 2.710 \\ 1.951 & 2.765 & 0.645 & 0.350 & 0.423 \\ 1.373 & 2.252 & 0.350 & 0.324 & 0.347 \\ 2.231 & 2.710 & 0.423 & 0.347 & 1.004 \end{bmatrix}$$

where, for example, the element in the second row and third column is

$$(20 \times 1.910 + 27 \times 3.398)/47 = 2.765$$

The inverse of the matrix **C** is found to be

$$\mathbf{C}^{-1} = \begin{bmatrix} 0.2061 & -0.0694 & -0.2395 & 0.0785 & -0.1969 \\ -0.0694 & 0.1234 & -0.0376 & -0.5517 & 0.0277 \\ -0.2395 & -0.0376 & 4.2219 & -3.2624 & -0.0181 \\ 0.0785 & -0.5517 & -3.2624 & 11.4610 & -1.2720 \\ -0.1969 & 0.0277 & -0.0181 & -1.2720 & 1.8068 \end{bmatrix}$$

This can be verified by evaluating the product $\mathbf{C} \times \mathbf{C}^{-1}$ and seeing that this is a unit matrix, apart from rounding errors.

Substituting the elements of $\mathbf{C}^{-1}$ and other values into Equation 4.7 produces

$$\begin{aligned} T^2 &= \{(21 \times 28)(21+28)\}\,[(157.381 - 158.429) \times 0.2061 \times (157.381 - 158.429) \\ &\quad - (157.318 - 158.429) \times 0.0694 \times (241.000 - 241.571) + \ldots \\ &\quad + (20.810 - 20.839) \times 1.8068 \times (20.810 - 20.839) \\ &= 2.824 \end{aligned}$$

Using Equation 4.8, this converts to an F-statistic of

$$F = (21 + 28 - 5 - 1) \times 2.824 / \{(21 + 28 - 2) \times 5\} = 0.517$$

with 5 and 43 df. Clearly, this is not significantly large because a significant F-value must exceed unity. Hence there is no evidence of a difference in population means for survivors and nonsurvivors, taking all five variables together.

4.4   *Multivariate versus univariate tests*

In the previous example, there were no significant results either for the variables considered individually or for the overall multivariate test. However, it is quite possible to have insignificant univariate tests but a significant multivariate test. This can occur because of the accumulation of the evidence from the individual variables in the overall test. Conversely, an insignificant multivariate test can occur when some univariate tests are significant. This can occur when the evidence of a difference provided by the significant variables is swamped by the evidence of no difference provided by the other variables.

One important aspect of the use of a multivariate test as distinct from a series of univariate tests concerns the control of type-one error rates. A type-one error involves finding a significant result when, in reality, the two samples being compared come from populations with the same mean (for a univariate test) or means (for a multivariate test). With a univariate test at the 5% level, there is a 0.95 probability of a nonsignificant result when the population means are the same. Hence if p independent tests are carried out under these conditions, then the probability of getting no significant results is 0.95^p. The probability of at least one significant result is therefore $1 - 0.95^p$, which may be unacceptably large. For example, if $p = 5$, then the probability of at least one significant result by chance alone is $1 - 0.95^5 = 0.23$. With multivariate data, variables are usually not independent, so $1 - 0.95^p$ does not quite give the correct probability of at least one significant result by chance alone if variables are tested one by one with univariate t-tests. However, the principle still applies that the more tests that are made, the higher is the probability of obtaining at least one significant result by chance.

On the other hand, a multivariate test such as Hotelling's T^2 test using the 5% level of significance gives a 0.05 probability of a type-one error, irrespective of the number of variables involved, providing that the assumptions of the test hold. This is a distinct advantage over a series of univariate tests, particularly when the number of variables is large.

There are ways of adjusting significance levels in order to control the overall probability of a type-one error when several univariate tests are carried out. The simplest approach involves using a Bonferroni adjustment. For example, if p univariate tests are carried out using the significance level $(5/p)\%$, then the probability of obtaining any significant results is 0.05 or less and α or less if the null hypothesis is true for each test. More generally, if p tests are carried out using the significance level $(100\alpha/p)\%$, then the probability of obtaining any significant results by chance is α or less.

Some people are not inclined to use a Bonferroni correction to significance levels because the significance levels applied to the individual tests become so extreme if p is large. For example, with $p = 10$ and an overall 5% level of significance, a univariate test result is declared significant only if it is significant at the 0.5% level. This has led to the development of some slightly less conservative variations on the Bonferroni correction, as discussed by Peres-Neto (1999) and Manly (2001, sec. 4.9).

It can certainly be argued that the use of a single multivariate test provides a better procedure in many cases than making a large number of univariate tests. A multivariate test also has the added advantage of taking proper account of the correlation between variables.

4.5 Comparison of variation for two samples: the single-variable case

With a single variable, the best-known method for comparing the variation in two samples is the F-test. If s_j^2 is the variance in the jth sample, calculated as shown in Equation 4.1, then the ratio s_1^2/s_2^2 is compared with percentage points of the F-distribution with $(n_1 - 1)$ and $(n_2 - 1)$ df. A value of the ratio that is significantly different from one is then evidence that the samples are from two populations with different variances. Unfortunately, the F-test is known to be rather sensitive to the assumption of normality. A significant result may well be due to the fact that a variable is not normally distributed rather than to unequal variances. For this reason, it is sometimes argued that the F-test should never be used to compare variances.

A robust alternative to the F-test is Levene's (1960) test. The idea here is to transform the original data in each sample into absolute deviations from the sample mean or the sample median, and then test for a significant difference between the mean deviations in the two samples using a t-test. Although absolute deviations from the sample means are sometimes used, a more robust test is likely to be obtained by using absolute deviations from the sample medians (Schultz, 1983). The procedure using medians is illustrated in Example 4.2 below.

4.6 Comparison of variation for two samples: the multivariate case

Many computer packages use Box's M-test to compare the variation in two or more multivariate samples. Because this applies for two or more samples, it is described in Section 4.8 below. This test is known to be rather sensitive to the assumption that the samples are from multivariate normal distributions. There is, therefore, always the possibility that a significant result is due to nonnormality rather than to unequal population covariance matrices.

An alternative procedure that should be more robust can be constructed using the principle behind Levene's test. This is done by transforming the data values into absolute deviations from sample means or medians. The question of whether two samples display significantly different amounts of variation is then transformed into a question of whether the transformed values show significantly different mean vectors. Testing of the mean vectors can be done using a T^2 test.

Another possibility was suggested by Van Valen (1978). This involves calculating

$$d_{ij} = \sqrt{\left\{ \sum_{k=1}^{P} \left(x_{ijk} - \bar{x}_{jk} \right)^2 \right\}} \qquad (4.10)$$

where x_{ijk} is the value of variable X_k for the ith individual in sample j, and $\bar{x}_{jk}$ is the mean of the same variable in the sample. The sample means of the d_{ij} values are compared with a t-test. Obviously, if one sample is more variable than another, then the mean d_{ij} values will tend to be higher in the more variable sample.

To ensure that all variables are given equal weight, they should be standardized before the calculation of the d_{ij} values. Coding them to have unit variances will achieve this. For a more robust test, it may be better to use sample medians in place of the sample means in Equation 4.10. In this case, the formula for d_{ij} values is

$$d_{ij} = \sqrt{\left\{ \sum_{k=1}^{P} \left(x_{ijk} - M_{jk} \right)^2 \right\}} \qquad (4.11)$$

where M_{jk} is the median for variable X_k in the jth sample.

The T^2 test and Van Valen's test for deviations from medians are illustrated in the example that follows. One point to note about the use of the test statistics (Equation 4.10 and Equation 4.11) is that they are based on an implicit assumption that if the two samples being tested differ, then one sample will be more variable than the other for all variables. A significant result cannot be expected in a case where, for example, X_1 and X_2 are more variable in sample 1 but X_3 and X_4 are more variable in sample 2. The effect of the differing variances would then tend to cancel out in the calculation of d_{ij}. Thus Van Valen's test is not appropriate for situations where changes in the level of variation are not expected to be consistent for all variables.

Example 4.2   Testing variation for female sparrows

With Bumpus's female sparrow data shown in Table 1.1, one of the interesting questions concerns whether the nonsurvivors were more variable than the survivors. This is what is expected if stabilizing selection took place.

To examine this question, first of all the individual variables can be considered one at a time, starting with X_1, the total length. For Levene's test, the original data values are transformed into deviations from sample medians. The median for survivors is 157 mm, and the absolute deviations from this median for the 21 birds in the sample then have a mean of $\bar{x}_1 = 2.57$ and a variance of $s_1^2 = 4.26$. The median for nonsurvivors is 159 mm, and the absolute deviations from this median for the 28 birds in the sample have a mean of $\bar{x}_2 = 3.29$ with a variance of $s_2^2 = 4.21$. The pooled variance from Equation 4.3 is 4.231, and the t-statistic of Equation 4.2 is

$$t = (2.57 - 3.29)\big/\left\{4.231\left(1/21 + 1/28\right)\right\}^{1/2} = -1.21$$

with 47 df.

Because nonsurvivors would be more variable than survivors if stabilizing selection occurred, it is a one-sided test that is required here, with low values of t providing evidence of selection. The observed value of t is not significantly low in the present instance. The t-values for the other variables are as follows: alar extent, $t = -1.18$; length of the beak and head, $t = -0.81$; length of the humerus, $t = -1.91$; and length of keel of the sternum, $t = -1.40$. Only for the length of the humerus is the result significantly low at the 5% level.

Table 4.2 shows the absolute deviations from sample medians for the data after they have been standardized for Van Valen's test. For example, the first value given for variable 1 (for survivors) is 0.28. This was obtained as follows. First, the original data were coded to have a zero mean and a unit variance for all 49 birds. This transformed the total length for the first survivor to $(156 - 157.98)/3.617 = -0.55$. The median transformed length for survivors was then -0.27. Hence the absolute deviation from the sample median for the first survivor is $|-0.55 - (-0.27)| = 0.28$, as recorded.

Comparing the transformed sample mean vectors for the five variables using Hotelling's T^2 test gives a test statistic of $T^2 = 4.75$, corresponding to an F-statistic of 0.87 with 5 and 43 df using Equation 4.8. There is therefore no evidence of a significant difference between the samples from this test because the F-value is less than one.

Finally, consider Van Valen's test. The d values from Equation 4.11 are shown in the last column of Table 4.2. The mean for survivors is 1.760, with variance 0.411. The mean for nonsurvivors is 2.265, with variance 1.133. The t-value from Equation 4.2 is then -1.92, which is significantly low at the 5% level. Hence this test indicates more variation for nonsurvivors than for survivors.

An explanation for the significant result with this test, but no significant result with Levene's test, is not hard to find. As noted above, Levene's test is not directional, and it does not take into account the expectation that the survivors will, if anything, be less variable than the nonsurvivors. On the other hand, Van Valen's test is specifically for less variation in sample 1 than in sample 2, for all variables. In the present case, all of the variables show less variation in sample 1 than in sample 2. Van Valen's test has emphasized this fact, but Levene's test has not.

Table 4.2 Absolute Deviations from Sample Medians for Bumpus's Data and d
Values from Equation 4.11

Total Length	Alar Extent	Beak and Head	Length of Humerus	Keel of Sternum	d
0.28	1.00	0.25	0.00	0.10	1.07
0.83	0.00	1.27	1.07	1.02	2.12
1.11	0.00	0.51	0.18	0.00	1.23
1.11	0.80	0.64	1.43	0.41	2.12
0.55	0.60	0.13	0.18	0.31	0.90
1.66	1.40	0.76	0.90	0.31	2.49
0.00	0.40	0.64	0.18	0.41	0.87
0.55	0.20	1.78	0.18	0.61	1.98
1.94	1.59	1.65	1.07	0.51	3.23
0.28	0.40	0.51	0.54	1.43	1.68
0.28	0.00	0.13	0.18	1.43	1.47
0.83	0.80	0.38	0.18	0.10	1.23
1.11	1.20	1.14	1.43	1.22	2.74
0.00	1.00	0.76	1.07	0.61	1.76
0.00	1.00	0.13	0.72	0.82	1.48
0.28	0.60	0.64	0.90	0.31	1.32
0.28	0.80	0.00	0.00	1.02	1.32
1.11	0.40	1.14	0.54	0.31	1.75
0.55	0.80	1.40	0.00	0.51	1.78
1.66	1.20	1.40	0.18	1.32	2.82
0.55	0.80	0.13	0.90	0.92	1.61
1.11	0.40	0.13	0.90	0.00	1.48
0.83	0.40	0.00	0.54	0.10	1.07
0.28	0.00	1.40	0.54	1.02	1.83
1.94	1.99	1.53	2.33	0.92	4.04
0.28	1.59	0.25	0.54	1.83	2.52
1.11	1.00	0.64	0.00	0.71	1.77
0.55	0.60	0.89	1.79	0.71	2.27
1.66	0.60	2.03	2.33	2.04	4.10
1.66	2.19	1.78	2.15	0.92	4.02
0.83	0.60	1.53	0.90	2.45	3.19
0.83	0.20	0.13	0.54	0.61	1.19
0.00	0.60	0.38	0.00	1.02	1.24
0.00	1.00	0.76	0.72	1.73	2.26
1.11	0.20	0.76	0.00	0.61	1.49
0.83	1.99	0.51	1.07	1.53	2.90
1.94	2.39	1.40	2.15	2.14	4.54
0.00	0.00	0.89	0.54	0.20	1.06
1.11	0.80	0.38	1.07	1.43	2.28
1.11	1.40	2.42	1.79	2.14	4.10
1.11	0.00	0.64	0.72	0.00	1.46
0.83	1.00	0.25	0.54	0.41	1.48
0.00	0.80	0.00	0.18	0.41	0.91
0.55	0.60	0.76	1.07	0.10	1.55
1.11	1.40	1.02	1.43	1.12	2.74
0.83	1.00	0.51	1.07	0.31	1.79
1.66	1.00	1.14	0.18	0.31	2.28
0.83	0.60	1.27	0.00	0.41	1.68
1.38	1.20	1.02	0.54	0.20	2.17

Table 4.3 One-Factor Analysis of Variance for Comparing the Mean Values of Samples from m Populations, with a Single Variable

Source of Variation	Sum of Squares	df	Mean Square	F-ratio
Between samples	$B = T - W$	$m - 1$	$M_B = B/(m-1)$	$F = M_B/M_W$
Within samples	$W = \sum\limits_{j=1}^{m} \sum\limits_{i=1}^{n_j} \left(x_{ij} - \bar{x}_j\right)^2$	$n - m$	$M_W = W/(n-m)$	
Total	$T = \sum\limits_{j=1}^{m} \sum\limits_{i=1}^{n_j} \left(x_{ij} - \bar{x}\right)^2$	$n - 1$		

Note: n_j = size of the jth sample; $n = n_1 + n_2 + \ldots + n_m$ = total number of observations; x_{ij} = ith observation in the jth sample; $\bar{x}_j$ = mean of the jth sample; $\bar{x}$ = mean of all observations.

4.7   *Comparison of means for several samples*

When there is a single variable and several samples to be compared, the generalization of the t-test is the F-test from a one-factor analysis of variance. The calculations are as shown in Table 4.3.

When there are several variables and several samples, the situation is complicated by the fact that there are four alternative statistics that are commonly used to test the hypothesis that all of the samples came from populations with the same mean vector.

The first test to be considered uses Wilks's lambda statistic

$$\Lambda = |\mathbf{W}| / |\mathbf{T}| \tag{4.12}$$

where $|\mathbf{W}|$ is the determinant of the within-sample sum of squares and cross-products matrix, and $|\mathbf{T}|$ is the determinant of the total sum of squares and cross-products matrix. Essentially, this compares the variation within the samples to the variation both within and between the samples. Here the matrices $\mathbf{T}$ and $\mathbf{W}$ require some further explanation. Let x_{ijk} denote the value of variable X_k for the ith individual in the jth sample; let $\bar{x}_{jk}$ denote the mean of X_k in the same sample; and let $\bar{x}_k$ denote the overall mean of X_k for all the data taken together. In addition, assume that there are m samples, with the jth of size n_j. Then the element in row r and column c of $\mathbf{T}$ is

$$t_{rc} = \sum\limits_{j=1}^{m} \sum\limits_{i=1}^{n_j} \left(x_{ijr} - \bar{x}_r\right)\left(x_{ijc} - \bar{x}_c\right) \tag{4.13}$$

and the element in row r and column c of $\mathbf{W}$ is

$$w_{rc} = \sum\limits_{j=1}^{m} \sum\limits_{i=1}^{n_j} \left(x_{ijr} - \bar{x}_{jr}\right)\left(x_{ijc} - \bar{x}_{jc}\right) \tag{4.14}$$

What is meant by a determinant is briefly discussed in Section 2.4. Here, all that needs to be known is that they are scalar quantities, i.e., ordinary numbers rather than vectors or matrices, and that special computer algorithms are needed to calculate them unless the matrices involved are of size 2×2 or, possibly, 3×3.

If Λ is small, then it indicates that the variation within the samples is low in comparison with the total variation. This provides evidence that the samples do not come from populations with the same mean vector. An approximate test for whether the within-sample variation is significantly low in this respect is described in Table 4.4. Tables of exact critical values are also available.

Let $\lambda_1 \geq \lambda_2 \geq \ldots \geq \lambda_p \geq 0$ be the eigenvalues of $\mathbf{W}^{-1}\mathbf{B}$, where $\mathbf{B} = \mathbf{T} - \mathbf{W}$ is called the between-sample matrix of sums of squares and cross-products, because the typical entry is the difference between a total sum of squares or cross-product minus the corresponding term within samples. Then Wilks's lambda can also be expressed as

$$\Lambda = \prod_{i=1}^{p} 1/(1+\lambda_i) \tag{4.15}$$

This is the form that is sometimes used to represent it.

A second statistic is the largest eigenvalue λ_1 of the matrix $\mathbf{W}^{-1}\mathbf{B}$, which leads to what is called Roy's largest root test (remembering that eigenvalues are also called latent roots). The basis for using this statistic is the result that if the linear combination of the variables X_1 to X_p that maximizes the ratio of the between-sample sum of squares to the within-sample sum of squares is found, then this maximum ratio equals λ_1. This then implies that this maximum eigenvalue should be a good statistic for testing whether the between-sample variation is significantly large, and that there is therefore evidence that the samples being considered do not come from populations with the same mean vector. This approach is related to discriminant function analysis, which is the subject of Chapter 8. It may be important to know that what some computer programs call Roy's largest root statistic is actually $\lambda_1/(1 - \lambda_1)$ rather than λ_1 itself. If in doubt, check the program documentation.

To assess whether λ_1 is significantly large, the exact probability of a value as large as the observed one can be calculated numerically, or an F-distribution can be used to find a lower bound to the significance level, i.e., the F-value is calculated and the true significance level is greater than the probability of obtaining a value this large or larger. Users of computer packages should be aware of which of these alternatives is used if a significant result is obtained. This is because if the F-distribution is used, then the value of λ_1 may not actually be significantly large at the chosen significance level. The F-value used is described in Table 4.4.

Table 4.4 Test Statistics Used To Compare Sample Mean Vectors with Approximate F-Tests for Evidence that the Population Values Are Not Constant

Test	Statistic	F	df_1	df_2	Comment
Wilks's lambda	Λ	$\{(1 - \Lambda^{1/t})/\Lambda^{1/t}\}\,(df_2/df_1)$	$p(m - 1)$	$wt - (df_1/2) + 1$	$w = n - 1 - \{(p + m)/2\}$ $t = [(df_1^2 - 4)/\{p^2 + (m - 1)^2 - 5\}]^{1/2}$ If $df_1 = 2$, set $t = 1$
Roy's largest root	λ_1	$(df_2/df_1)\,\lambda_1$	d	$n - m - d - 1$	The significance level obtained is a lower bound $d = \max(p, m - 1)$. $s = \min(p, m - 1) =$ number of
Pillai's trace	$V = \sum_{i=1}^{p} \lambda_i/(1 + \lambda_i)$	$(n - m - p + s)\,V/\{d\,(s - V)\}$	sd	$s(n - m - p + s)$	positive eigenvalues $d = \max(p, m - 1)$. s is as for Pillai's trace
Lawes–Hotelling trace	$U = \sum_{i=1}^{p} \lambda_i$	$df_2\,U/(s\,df_1)$	$s(2A + s + 1)$	$2(sB + 1)$	$A = (\lvert m - p - 1 \rvert - 1)/2$ $B = (n - m - p - 1)/2$

Note: It is assumed that there are p variables in m samples, with the ith of size n_i, and a total sample size of $n = \Sigma n_i$. These are approximations for general p and m. Exact or better approximations are available for some special cases, and other approximations are also available. In all cases, the test statistic is transformed to the stated F-value, and this is tested to see whether it is significantly large in comparison with the F-distribution with df_1 and df_2 degrees of freedom. Chi-squared distribution approximations are also in common use, and tables of critical values are available (Kres, 1983).

The third statistic often used to test whether the samples come from populations with the same mean vectors is Pillai's trace statistic This can be written in terms of the eigenvalues λ_1 to λ_p as

$$V = \sum_{i=1}^{P} \lambda_i / (1 + \lambda_i) \qquad (4.16)$$

Again, large values for this statistic provide evidence that the samples being considered come from populations with different mean vectors. An approximation of the significance level (the probability of obtaining a value as large or larger than V if the samples are from populations with the same mean vector) is again provided in Table 4.4.

Finally, the fourth statistic often used to test the null hypothesis of equal population mean vectors is the Lawes–Hotelling trace

$$U = \sum_{i=1}^{P} \lambda_i \qquad (4.17)$$

which is just the sum of the eigenvalues of the matrix $\mathbf{W}^{-1}\,\mathbf{B}$. Yet again, large values provide evidence against the null hypothesis, with an approximate F-test provided in Table 4.4.

Generally, the four tests just described can be expected to give similar significance levels, so there is no real need to choose between them. They all involve the assumption that the distribution of the p variables is multivariate normal, with the same within-sample covariance matrix for all of the m populations that the samples are drawn from. They are all also considered to be fairly robust if the sample sizes are equal or nearly so for the m samples. If there are questions about either the multivariate normality or the equality of covariance matrices, then simulation studies suggest that Pillai's trace statistic may be more robust that the other three statistics (Seber, 1984, p. 442).

4.8   Comparison of variation for several samples

Box's M-test is the best known for comparing the variation in several samples. This test has already been mentioned for the two-sample situations with several variables to be compared, and it can be used with one or several variables, with two or more samples.

For m samples, the M-statistic is given by the equation

$$M = \left\{ \prod_{i=1}^{m} |\mathbf{C}_i|^{(n_i-1)/2} \right\} \Big/ |\mathbf{C}|^{(n-m)/2} \qquad (4.18)$$

where n_i is the size of the ith sample, C_i is the sample covariance for the ith sample as defined in Section 2.7, C is the pooled covariance matrix

$$C = \sum_{i=1}^{m} (n_i - 1)\, C_i / (n - m)$$

and $n = \Sigma\, n_i$ is the total number of observations.

Large values of M provide evidence that the samples are not from populations with the same covariance matrix. An approximate F-test for whether an observed M-value is significantly large is provided by calculating

$$F = -2\, b \log_e (M) \tag{4.19}$$

and finding the probability of a value this large or larger for an F-distribution with v_1 and v_2 degrees of freedom, where

$$v_1 = p(p+1)(m-1)/2$$
$$v_2 = (v_1 + 2)/(c_2 - c_1^2)$$

and

$$b = (1 - c_1 - v_1/v_2)/v_1$$

where

$$c_1 = (2p^2 + 3p - 1)\left\{\sum_{i=1}^{m} 1/(n_i - 1) - 1/(n - m)\right\} \Big/ \left\{6(p+1)(m-1)\right\}$$

and

$$c_2 = (p-1)(p+2)\left\{\sum_{i=1}^{m} 1/(n_i - 1)^2 - 1/(n - m)^2\right\} \Big/ \left\{6(m-1)\right\}$$

The F-approximation of Equation 4.19 is only valid for $c_2 > c_1^2$. If $c_2 < c_1^2$, then an alternative approximation is used. In this alternative case, the F-value is calculated to be

$$F = \left\{2\, b_1 v_2 \log_e (M)\right\} / \left\{v_1 + 2\, b_1 \log_e (M)\right\} \tag{4.20}$$

where

$$b_1 = \left(1 - c_1 - 2/v_2\right)/v_2$$

This is tested against the F-distribution with v_1 and v_2 df to see if it is significantly large.

Box's test is known to be sensitive to deviations from normality in the distribution of the variables being considered. For this reason, robust alternatives to Box's test are recommended here, these being generalizations of what was suggested for the two-sample situation. Thus absolute deviations from sample medians can be calculated for the data in m samples. For a single variable, these can be treated as the observations for a one-factor analysis of variance. A significant F-ratio is then evidence that the samples come from populations with different mean deviations, i.e., populations with different covariance matrices. With more than one variable, any of the four tests described in the last section can be applied to the transformed data, and a significant result indicates that the covariance matrix is not constant for the m populations sampled.

Alternatively, the variables can be standardized to have unit variances for all the data lumped together, and d values can be calculated using Equation 4.11. These d values can then be analyzed by a one-factor analysis of variance. This generalizes Van Valen's test, which was suggested for comparing the variation in two multivariate samples. A significant F-ratio from the analysis of variance indicates that some of the m populations sampled are more variable than others. As in the two-sample situation, this test is only really appropriate when some samples may be more variable than others for all the measurements being considered.

Example 4.3   Comparison of samples of Egyptian skulls

As an example of the tests for comparing several samples, consider the data shown in Table 1.2 for four measurements on male Egyptian skulls for five samples from various past ages.

A one-factor analysis of variance on the first variable, maximum breadth, provides F = 5.95, with 4 and 145 df (Table 4.3). This is significantly large at the 0.1% level, and hence there is clear evidence that the population mean changed with time. For the other three variables, analysis of variance provides the following results: basibregmatic height, F = 2.45 (significant at the 5% level); basialveolar length, F = 8.31 (significant at the 0.1% level); and nasal height, F = 1.51 (not significant). Hence, there is evidence that the population mean changed with time for the first three variables.

Next, consider the four variables together. If the five samples are combined, then the matrix of sums of squares and products for the 150 observations, calculated using Equation 4.13, is

$$T = \begin{bmatrix} 3563.89 & -222.81 & -615.16 & 426.73 \\ -222.81 & 3635.17 & 1046.28 & 346.47 \\ -615.16 & 1046.28 & 4309.27 & -16.40 \\ 426.73 & 346.47 & -16.40 & 1533.33 \end{bmatrix}$$

for which the determinant is $|T| = 7.306 \times 10^{13}$. Also, the within-sample matrix of sums of squares and cross-products is found from Equation 4.14 to be

$$W = \begin{bmatrix} 3061.07 & 5.33 & 11.47 & 291.30 \\ 5.33 & 3405.27 & 754.00 & 412.53 \\ 11.47 & 754.00 & 3505.97 & 164.33 \\ 291.30 & 412.53 & 164.33 & 1472.13 \end{bmatrix}$$

for which the determinant is $|W| = 4.848 \times 10^{13}$. Wilks's lambda statistic is therefore

$$\Lambda = |W| / |T| = 0.6636$$

The details of an approximate F-test to assess whether this value is signifi-cantly small are provided in Table 4.4. With $p = 4$ variables, $m = 5$ samples, and $n = 150$ observations in total, it is found using the notation in Table 4.4 that

$$df_1 = p(m-1) = 16,$$
$$w = n - 1 - (p+m)/2 = 150 - 1 - (4+5)/2 = 144.5,$$
$$t = \left[(df_1^2 - 4) / \left\{ p^2 + (m-1)^2 - 5 \right\} \right]^{1/2} = \left[(16^2 - 4) / \left\{ 4^2 + (5-1)^2 - 5 \right\} \right]^{1/2} = 3.055$$

and

$$df_2 = wt - df_1/2 + 1 = 144.5 \times 3.055 - 16/2 + 1 = 434.5$$

The F-statistic is then

$$F = \left\{ (1 - \Lambda^{1/t}) / \Lambda^{1/t} \right\} (df_2/df_1) = \left\{ (1 - 0.6636^{1/3.055}) / 0.6636^{1/3.055} \right\} (434.5/16) = 3.90$$

with 16 and 434.5 df. This is significantly large at the 0.1% level ($p < 0.001$). There is therefore clear evidence that the vector of mean values of the four variables changed with time.

The maximum root of the matrix $\mathbf{W}^{-1}\mathbf{B}$ is $_1 = 0.4251$ for Roy's maximum root test. The corresponding approximate F-statistic from Table 4.4 is

$$F = (df_2/df_1)\,\lambda_1 = (140/4)\,0.4251 = 14.88$$

with 4 and 140 df, using the equations given in Table 4.4 for the df. This again is very significantly large (p < 0.001).

Pillai's trace statistic is V = 0.3533. The approximate F-statistic in this case is

$$F = (n - m - p + s)\,V/\{d(s - V)\} = 3.51$$

with sd = 16 and s (n – m – p + s) = 580 df, using the equations given in Table 4.4. This is another very significant result (p < 0.001).

Finally, for the tests on the mean vectors, the Lawley-Hotelling trace statistic has the value U = 0.4818. It is found using the equations in Table 4.4 that the intermediate quantities that are needed are s = 4, A = –0.5, and B = 70, so that the df values for the F-statistic are df_1 = s(2A + s + 1) = 16 and df_2 = 2(sB + 1) = 562. The F-statistic is then

$$F = df_2\,U/(s\,df_1) = (562 \times 0.4818)/(4 \times 16) = 4.23$$

Yet again, this is a very significant result (p < 0.001).

To compare the variation in the five samples, first consider Box's test. Equation 4.18 gives $M = 2.869 \times 10^{-11}$. The equations in the previous section (following Equation 4.19) then give b = 0.0235, and

$$F = -2\,b\,\log_e(M) = 1.14$$

with v_1 = 40 and v_2 = 46,379 df. This is not at all significantly large (p = 0.250), so this test gives no evidence that the covariance matrix changed with time.

Box's test is reasonable with this set of data because body measurements tend to have distributions that are close to normal. However, robust tests can also be carried out. It is a straightforward matter to transform the data into absolute deviations from sample medians for Levene-type tests. Analysis of variance then shows no significant difference between the sample means of the transformed data for any of the four variables considered individually. Also, none of the multivariate tests summarized in Table 4.4 give a result that is anything like significant at the 5% level for all the transformed variables taken together.

It appears, therefore, that although there is very strong evidence that mean values changed with time for the four variables being considered, there is no evidence that the variation changed.

4.9 Computer programs

The tests for multivariate normal data that are discussed in this chapter are fairly readily available in standard statistical computer packages, although many packages will be missing one or two of them. Also, the results of tests based on F-distribution approximations may vary to some extent from one program to the next because of the use of different approximations. On the other hand, for the robust tests on variances, some or all of the calculations may need to be done in a spreadsheet.

This chapter has been restricted to situations where there are two or more multivariate samples being compared to see whether they seem to come from populations with different mean vectors or from populations with different covariance matrices. In terms of mean vectors, this is the simplest case of what is sometimes called multivariate analysis of variance, or MANOVA for short. More complicated examples involve samples being classified on the basis of several factors, giving a generalization of ordinary analysis of variance (ANOVA). Many statistical packages allow the general MANOVA calculations to be performed.

4.10 Chapter summary

- The t-test is described for comparing the mean values of two samples from populations with equal variances. Welch's test is also described for situations where the population variances are not equal. Both tests assume that the data values are normally distributed.
- Hotelling's T^2 test is described as the generalization of the t-test when there is more than one variable being measured for two samples.
- The multiple testing problem (i.e., the type-one error rate becoming inflated if several univariate tests are carried out at one time) is discussed, together with the advantage of using one multivariate test instead.
- The F-test and Levene's robust test are described for comparing the variation in two samples.
- Box's M-test, a robust alternative based on the approach used in Levene's test, and Van Valen's test are described for comparing the variation in two samples.
- Four tests are described for comparing the mean vectors of several multivariate samples. These are Wilks's lambda test, Roy's largest root test, Pillai's trace statistic, and the Lawes–Hotelling trace. Approximate F-tests are described for assessing the significance of the statistics involved in each case.
- Box's M-test is described as a means for comparing the variation in several multivariate samples. This is noted to be very sensitive to the required assumption that the data are multivariate normally distributed. Robust alternatives using the principle behind Levene's test are therefore described.

- Multivariate analysis of variance (MANOVA) is mentioned as a generalization of the methods discussed in this chapter.

Exercise

Example 1.4 concerned the comparison between prehistoric dogs from Thailand and six other related animal groups in terms of mean mandible measurements. Table 4.5 shows some further data for the comparison of these groups that are part of the more extensive data discussed in the paper by Higham et al. (1980).

1. Test for significant differences between the five species in terms of the mean values and the variation in the nine variables. Test both for overall differences and for differences between the prehistoric Thai dogs and each of the other groups singly. What conclusion do you draw with regard to the similarity between prehistoric Thai dogs and the other groups?
2. Is there evidence of differences between the size of males and females of the same species for the first four groups?
3. Using a suitable graphical method, compare the distribution of the nine variables for the prehistoric and modern Thai dogs.

Table 4.5 Values for Nine Mandible Measurements for Samples of Five Canine Groups (mm)

	X_1	X_2	X_3	X_4	X_5	X_6	X_7	X_8	X_9	Sex[a]
				Modern Dogs from Thailand						
1	123	10.1	23	23	19	7.8	32	33	5.6	1
2	137	9.6	19	22	19	7.8	32	40	5.8	1
3	121	10.2	18	21	21	7.9	35	38	6.2	1
4	130	10.7	24	22	20	7.9	32	37	5.9	1
5	149	12.0	25	25	21	8.4	35	43	6.6	1
6	125	9.5	23	20	20	7.8	33	37	6.3	1
7	126	9.1	20	22	19	7.5	32	35	5.5	1
8	125	9.7	19	19	19	7.5	32	37	6.2	1
9	121	9.6	22	20	18	7.6	31	35	5.3	2
10	122	8.9	20	20	19	7.6	31	35	5.7	2
11	115	9.3	19	19	20	7.8	33	34	6.5	2
12	112	9.1	19	20	19	6.6	30	33	5.1	2
13	124	9.3	21	21	18	7.1	30	36	5.5	2
14	128	9.6	22	21	19	7.5	32	38	5.8	2
15	130	8.4	23	20	19	7.3	31	40	5.8	2
16	127	10.5	25	23	20	8.7	32	35	6.1	2

(continued)

Table 4.5 (continued)   Values for Nine Mandible Measurements for Samples of
Five Canine Groups (mm)

	X_1	X_2	X_3	X_4	X_5	X_6	X_7	X_8	X_9	Sex[a]
					Golden Jackals					
1	120	8.2	18	17	18	7.0	32	35	5.2	1
2	107	7.9	17	17	20	7.0	32	34	5.3	1
3	110	8.1	18	16	19	7.1	31	32	4.7	1
4	116	8.5	20	18	18	7.1	32	33	4.7	1
5	114	8.2	19	18	19	7.9	32	33	5.1	1
6	111	8.5	19	16	18	7.1	30	33	5.0	1
7	113	8.5	17	18	19	7.1	30	34	4.6	1
8	117	8.7	20	17	18	7.0	30	34	5.2	1
9	114	9.4	21	19	19	7.5	31	35	5.3	1
10	112	8.2	19	17	19	6.8	30	34	5.1	1
11	110	8.5	18	17	19	7.0	31	33	4.9	2
12	111	7.7	20	18	18	6.7	30	32	4.5	2
13	107	7.2	17	16	17	6.0	28	35	4.7	2
14	108	8.2	18	16	17	6.5	29	33	4.8	2
15	110	7.3	19	15	17	6.1	30	33	4.5	2
16	105	8.3	19	17	17	6.5	29	32	4.5	2
17	107	8.4	18	17	18	6.2	29	31	4.3	2
18	106	7.8	19	18	18	6.2	31	32	4.4	2
19	111	8.4	17	16	18	7.0	30	34	4.7	2
20	111	7.6	19	17	18	6.5	30	35	4.6	2
					Cuons					
1	123	9.7	22	21	20	7.8	27	36	6.1	1
2	135	11.8	25	21	23	8.9	31	38	7.1	1
3	138	11.4	25	25	22	9.0	30	38	7.3	1
4	141	10.8	26	25	21	8.1	29	39	6.6	1
5	135	11.2	25	25	21	8.5	29	39	6.7	1
6	136	11.0	22	24	22	8.1	31	39	6.8	1
7	131	10.4	23	23	23	8.7	30	36	6.8	1
8	137	10.6	25	24	21	8.3	28	38	6.5	1
9	135	10.5	25	25	21	8.4	29	39	6.9	1
10	131	10.9	25	24	21	8.5	29	35	6.2	2
11	130	11.3	22	23	21	8.7	29	37	7.0	2
12	144	10.8	24	26	22	8.9	30	42	7.1	2
13	139	10.9	26	23	22	8.7	30	39	6.9	2
14	123	9.8	23	22	20	8.1	26	34	5.6	2
15	137	11.3	27	26	23	8.7	30	39	6.5	2
16	128	10.0	22	23	22	8.7	29	37	6.6	2
17	122	9.9	22	22	20	8.2	26	36	5.7	2

(continued)

Table 4.5 (continued)   Values for Nine Mandible Measurements for Samples of Five Canine Groups (mm)

	X_1	X_2	X_3	X_4	X_5	X_6	X_7	X_8	X_9	Sex[a]
					Indian Wolves					
1	167	11.5	29	28	25	9.5	41	45	7.2	1
2	164	12.3	27	26	25	10.0	42	47	7.9	1
3	150	11.5	21	24	25	9.3	41	46	8.5	1
4	145	11.3	28	24	24	9.2	36	41	7.2	1
5	177	12.4	31	27	27	10.5	43	50	7.9	1
6	166	13.4	32	27	26	9.5	40	47	7.3	1
7	164	12.1	27	24	25	9.9	42	45	8.3	1
8	165	12.6	30	26	25	7.7	40	43	7.9	1
9	131	11.8	20	24	23	8.8	38	40	6.5	2
10	163	10.8	27	24	24	9.2	39	48	7.0	2
11	164	10.7	24	23	26	9.5	43	47	7.6	2
12	141	10.4	20	23	23	8.9	38	43	6.0	2
13	148	10.6	26	21	24	8.9	39	40	7.0	2
14	158	10.7	25	25	24	9.8	41	45	7.4	2
					Prehistoric Thai Dogs					
1	112	10.1	17	18	19	7.7	31	33	5.8	0
2	115	10.0	18	23	20	7.8	33	36	6.0	0
3	136	11.9	22	25	21	8.5	36	39	7.0	0
4	111	9.9	19	20	18	7.3	29	34	5.3	0
5	130	11.2	23	27	20	9.1	35	35	6.6	0
6	125	10.7	19	26	20	8.4	33	37	6.3	0
7	132	9.6	19	20	19	9.7	35	38	6.6	0
8	121	10.7	21	23	19	7.9	32	35	6.0	0
9	122	9.8	22	23	18	7.9	32	35	6.1	0
10	124	9.5	20	24	19	7.6	32	37	6.0	0

Note: The variables are X_1 = length of mandible; X_2 = breadth of mandible below first molar; X_3 = breadth of articular condyle; X_4 = height of mandible below first molar; X_5 = length of first molar; X_6 = breadth of first molar; X_7 = length of first to third molar, inclusive (first to second for cuon); X_8 = length from first to fourth premolar, inclusive; and X_9 = breadth of lower canine.

[a] Sex code is 1 for male, 2 for female, and 0 for unknown.

References

Carter, E.M., Khatri, C.G., and Srivastava, M.S. (1979), The effect of inequality of variances on the t-test, *Sankhya*, 41, 216–225.

Higham, C.F.W., Kijngam, A., and Manly, B.F.J. (1980), An analysis of prehistoric canid remains from Thailand, *J. Archaeological Sci.*, 7: 149–165.

Kres, H. (1983), *Statistical Tables for Multivariate Analysis*, Springer-Verlag, New York.

Levene, H. (1960), Robust tests for equality of variance, in *Contributions to Probability and Statistics*, Olkin, I., Ghurye, S.G., Hoeffding, W., Madow, W.G., and Mann, H.B., Eds., Stanford University Press, Stanford, CA, pp. 278–292.

Manly, B.F.J. (2001), *Statistics for Environmental Science and Management*, Chapman and Hall/CRC, Boca Raton, FL.

Manly, B.F.J. and Francis, R.I.C.C. (2002), Testing for mean and variance differences with samples from distributions that may be non-normal with unequal variances, *J. Statistical Computation Simulation*, 72, 633–646.

Peres-Neto, P.R. (1999), How many tests are too many? The problem of conducting multiple ecological inferences revisited, *Mar. Ecol. Prog. Ser.*, 176, 303–306.

Schultz, B. (1983), On Levene's test and other statistics of variation, *Evolutionary Theory*, 6, 197–203.

Seber, G.A.F. (1984), *Multivariate Observations*, Wiley, New York.

Van Valen, L. (1978), The statistics of variation, *Evolutionary Theory*, 4, 33–43 (erratum, *Evolutionary Theory*, 4, 202.)

Welch, B.L. (1951), On the comparison of several mean values: an alternative approach, *Biometrika*, 38, 330–336.

Yao, Y. (1965), An approximate degrees of freedom solution to the multivariate Behrens–Fisher problem, *Biometrika*, 52, 139–147.

chapter five

Measuring and testing multivariate distances

5.1 Multivariate distances

A large number of multivariate problems can be viewed in terms of distances between single observations, between samples of observations, or between populations of observations. For example, considering the data in Table 1.4 on mandible measurements of dogs, wolves, jackals, cuons, and dingos, it is sensible to ask how far one of these groups is from the other six groups. The idea then is that if two animals have similar mean mandible measurements, then they are close, whereas if they have rather different mean measurements, then they are distant from each other. Throughout this chapter, it is this concept of distance that is used.

A large number of distance measures have been proposed and used in multivariate analyses. Only some of the most common ones will be mentioned here. It is fair to say that measuring distances is a topic where a certain amount of arbitrariness seems unavoidable.

One situation is that there are n objects being considered, with a number of measurements being taken on each of these, and the measurements are of two types. For example, in Table 1.3, results are given for four environmental variables and six gene frequencies for 16 colonies of a butterfly species. Two sets of distances, environmental and genetic, can therefore be calculated between the colonies. An interesting question is then whether there is a significant relationship between these two sets of distances. Mantel's test (Section 5.6) is useful in this context.

5.2 Distances between individual observations

To begin with, consider the simplest case where there are n objects, each of which has values for p variables, $X_1, X_2, \ldots, X_p$. The values for object i can then be denoted by $x_{i1}, x_{12}, \ldots, x_{ip}$, and those for object j by $x_{j1}, x_{j2}, \ldots, x_{jp}$. The problem is to measure the distance between these two objects. If there are

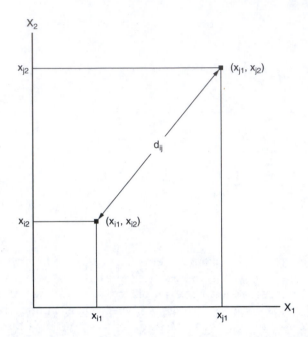

Figure 5.1 The Euclidean distance between objects i and j with p = 2 variables.

only p = 2 variables, then the values can be plotted as shown in Figure 5.1. Pythagoras's theorem then says that the length d_{ij} of the line joining the point for object i to the point for object j (the Euclidean distance) is

$$d_{ij} = \left\{ \left(x_{i1} - x_{j1} \right)^2 + \left(x_{i2} - x_{j2} \right)^2 \right\}^{1/2}$$

With p = 3 variables, the values can be taken as the coordinates in space for plotting the positions of individuals i and j (Figure 5.2). Pythagoras's theorem then gives the distance between the two points to be

$$d_{ij} = \left\{ \left(x_{i1} - x_{j1} \right)^2 + \left(x_{i2} - x_{j2} \right)^2 + \left(x_{i3} - x_{j3} \right)^2 \right\}^{1/2}$$

With more than three variables, it is not possible to use variable values as the coordinates for physically plotting points. However, the two- and three-variable cases suggest that the generalized Euclidean distance

$$d_{ij} = \left\{ \sum_{k=1}^{p} \left(x_{ik} - x_{jk} \right)^2 \right\} \qquad (5.1)$$

may serve as a satisfactory measure for many purposes with p variables.

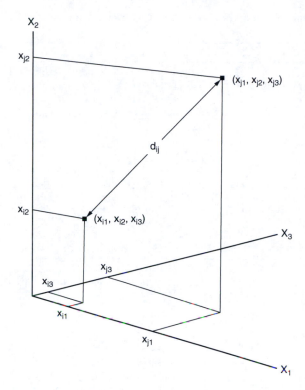

Figure 5.2 The Euclidean distance between objects i and j with p = 3 variables.

From the form of Equation 5.1, it is clear that if one of the variables measured is much more variable than the others, then this will dominate the calculation of distances. For example, to take an extreme case, suppose that n men are being compared, and that X_1 is their stature and the other variables are tooth dimensions, with all the measurements being in millimeters. Stature differences will then be in the order of perhaps 20 or 30 mm, while tooth dimension differences will be in the order of 1 or 2 mm. The simple calculations of d_{ij} will then provide distances between individuals that are essentially stature differences only, with tooth differences having negligible effects.

In practice, it is usually desirable for all variables to have about the same influence on the distance calculation. This can be achieved by a preliminary scaling by dividing each variable by its standard deviation for the n individuals being compared.

Example 5.1   Distances between dogs and related species

Consider again the data in Table 1.4 for mean mandible measurements of seven groups of Thai dogs and related species. It may be recalled from Chapter 1 that the main question with these data is how the prehistoric Thai dogs relate to the other groups.

Table 5.1 Standardized Variable Values Calculated from the Original Data in Table 1.4

Group	X_1	X_2	X_3	X_4	X_5	X_6
Modern dog	−0.46	−0.46	−0.68	−0.69	−0.45	−0.57
Golden jackal	−1.41	−1.79	−1.04	−1.29	−0.80	−1.21
Chinese wolf	1.78	1.48	1.70	1.80	1.55	1.50
Indian wolf	0.60	0.55	0.96	0.69	1.17	0.88
Cuon	0.13	0.31	−0.04	0.00	−1.10	−0.37
Dingo	−0.52	0.03	−0.13	−0.17	0.03	0.61
Prehistoric dog	−0.11	−0.12	−0.78	−0.34	−0.41	−0.83

Note: X_1 = breadth of mandible; X_2 = height of mandible below the first molar; X_3 = length of the first molar; X_4 = breadth of the first molar; X_5 = length from first to third molar, inclusive; and X_6 = length from first to fourth premolar, inclusive.

Table 5.2 Euclidean Distances between Seven Canine Groups

	Modern dog	Golden jackal	Chinese wolf	Indian wolf	Cuon	Dingo	Prehistoric dog
Modern dog	—						
Golden jackal	1.91	—					
Chinese wolf	5.38	7.12	—				
Indian wolf	3.38	5.06	2.14	—			
Cuon	1.51	3.19	4.57	2.91	—		
Dingo	1.56	3.18	4.21	2.20	1.67	—	
Prehistoric dog	0.66	2.39	5.12	3.24	1.26	1.71	—

The first step in calculating distances is to standardize the measurements. Here this will be done by expressing them as deviations from means in units of standard deviations. For example, the first measurement X_1 (breadth) has a mean of 10.486 mm and a sample standard deviation of 1.697 mm for the seven groups. The standardized variable values are then calculated as: modern dog, $(9.7 − 10.486)/1.697 = −0.46$; golden jackal, $(8.1 − 10.486)/1.697 = −1.41$; prehistoric dog, $(10.3 − 10.486)/1.697 = −0.11$; and so on for the other groups of dogs. Standardized values for all the variables are shown in Table 5.1.

Using Equation 5.1, the distances shown in Table 5.2 have been calculated from the standardized variables. It is clear that the prehistoric dogs are rather similar to modern dogs in Thailand because the distance between these two groups (0.66) is by far the smallest distance in the whole table. Higham et al. (1980) concluded from a more complicated analysis that the modern and prehistoric dogs are indistinguishable.

5.3   *Distances between populations and samples*

A number of measures have been proposed for the distance between multivariate populations when information is available on the means, variances, and covariances of the populations. Two measures will be considered here.

Suppose that two or more populations are available, and the multivariate distributions in these populations are known for p variables $X_1, X_2, \ldots, X_p$. Let the mean of variable X_k in the ith population be μ_{ki}, and assume that the variance of X_k is V_k in all the populations. Then Penrose (1953) proposed the relatively simple measure

$$P_{ij} = \sum_{k=1}^{P} \left(\mu_{ki} - \mu_{kj}\right)^2 \Big/ \left(pV_k\right) \tag{5.2}$$

for the distance between population i and population j.

A disadvantage of Penrose's measure is that it does not take into account the correlations between the p variables. This means that when two variables are measuring essentially the same thing, and hence are highly correlated, they still individually both contribute about the same amount to population distances as a third variable that is uncorrelated with all other variables.

A measure that does take into account the correlations between variables is the Mahalanobis (1948) distance

$$D_{ij}^2 = \sum_{r=1}^{P} \sum_{s=1}^{P} \left(\mu_{ri} - \mu_{rj}\right) v^{rs} \left(\mu_{si} - \mu_{sj}\right) \tag{5.3}$$

where v^{rs} is the element in the rth row and sth column of the inverse of the population covariance matrix for the p variables. This is a quadratic form that can also be written as

$$D_{ij}^2 = \left(\boldsymbol{\mu}_i - \boldsymbol{\mu}_j\right)' \mathbf{V}^{-1} \left(\boldsymbol{\mu}_i - \boldsymbol{\mu}_j\right)$$

where $\boldsymbol{\mu}_i$ is the population mean vector for the ith population, and $\mathbf{V}$ is the population covariance matrix. This measure requires the assumption that $\mathbf{V}$ is the same for all populations.

A Mahalanobis distance is also often used to measure the distance of a single multivariate observation from the center of the population that the observation comes from. If $x_1, x_2, \ldots, x_p$ are the values of $X_1, X_2, \ldots, X_p$ for the individual, with corresponding population mean values of $\mu_1, \mu_2, \ldots, \mu_p$, then this distance is

$$D^2 = \sum_{r=1}^{P} \sum_{s=1}^{P} \left(x_r - \mu_r\right) v^{rs} \left(x_s - \mu_s\right) \tag{5.4}$$

$$= \left(\mathbf{x} - \boldsymbol{\mu}\right)' \mathbf{V}^{-1} \left(\mathbf{x} - \boldsymbol{\mu}\right)$$

where $\mathbf{x} = (x_1, x_2, \ldots, x_p)$, $\boldsymbol{\mu}$ is the population mean vector, $\mathbf{V}$ is the population covariance matrix, and as before, v^{rs} is the element in the rth row and sth column of the inverse of $\mathbf{V}$.

The value of D^2 can be thought of as a multivariate residual for the observation x, i.e., a measure of how far the observation x is from the center of the distributions of all values, taking into account all the variables being considered and their covariances. An important result is that if the population being considered is multivariate normally distributed, then the values of D^2 will follow a chi-squared distribution with p degrees of freedom (df) if x comes from this distribution. A significantly large value of D^2 means that the corresponding observation is either (a) a genuine but unlikely record, (b) an observation from another distribution, or (c) a record containing some mistake. Observations with large Mahalanobis residuals should therefore be examined to see whether they have just been recorded wrongly.

Equation 5.2 to Equation 5.4 can be used with sample data if estimates of population means, variances, and covariances are used in place of true values. In that case, the covariance matrix **V** involved in Equation 5.3 and Equation 5.4 should be replaced with the pooled estimate from all the samples available, as defined in Section 4.8 for Box's M-test.

In principle, the Mahalanobis distance is superior to the Penrose distance because it uses information on covariances. However, this advantage is only present when the covariances are accurately known. When covariances can only be estimated rather poorly from small samples, it is probably best to use the simpler Penrose measure. It is difficult to say precisely what a small sample means in this context. Certainly there should be no problem with using Mahalanobis distances based on a covariance matrix estimated with a total sample size of 100 or more.

Example 5.2   Distances between samples of Egyptian skulls

For the data for the five samples of male Egyptian skulls shown in Table 1.2, the mean vectors and covariance matrices are shown in Table 5.3, as is the pooled covariance matrix. Although the five sample covariance matrices appear to differ somewhat, it has been shown in Example 4.3 that the differences are not significant.

The equation for Penrose's distance measures (Equation 5.2) can now be calculated between each pair of samples. There are p = 4 variables with variances that are estimated by $V_1 = 21.112$, $V_2 = 23.486$, $V_3 = 24.180$, and $V_4 = 10.154$, these being the diagonal terms in the pooled covariance matrix (Table 5.3). The sample mean values given in the vectors $\bar{x}_1$ to $\bar{x}_5$ are estimates of population means. For example, the distance between sample 1 and sample 2 is calculated as

$$P_{12} = (131.37 - 132.37)^2 / (4 \times 21.112) + (133.60 - 132.70)^2 / (4 \times 23.486)$$
$$+ (99.17 - 99.07)^2 / (4 \times 24.180) + (50.53 - 50.23)^2 / (4 \times 10.154)$$
$$= 0.023.$$

Table 5.3 The Samples' Mean Vectors and Covariance Matrices, and the Pooled Sample Covariance Matrix for the Egyptian Skull Data

Sample		Mean Vector	Sample Covariance Matrices			
			X_1	X_2	X_3	X_4
1	X_1	131.37	26.31	4.15	0.45	7.25
	X_2	133.60	4.15	19.97	−0.79	0.39
	X_3	99.17	0.45	−0.79	34.63	−1.92
	X_4	50.53	7.25	0.39	−1.92	7.64
2	X_1	132.37	23.14	1.01	4.77	1.84
	X_2	132.70	1.01	21.60	3.37	5.62
	X_3	99.07	4.77	3.37	18.89	0.19
	X_4	50.23	1.84	5.62	0.19	8.74
3	X_1	134.47	12.12	0.79	−0.78	0.90
	X_2	133.80	0.79	24.79	3.59	−0.09
	X_3	96.03	−0.78	3.59	20.72	1.67
	X_4	50.57	0.90	−0.09	1.67	12.60
4	X_1	135.50	15.36	−5.53	−2.17	2.05
	X_2	132.30	−5.53	26.36	8.11	6.15
	X_3	94.53	−2.17	8.11	21.09	5.33
	X_4	51.97	2.05	6.15	5.33	7.96
5	X_1	136.17	28.63	−0.23	−1.88	−1.99
	X_2	130.33	−0.23	24.71	11.72	2.15
	X_3	93.50	−1.88	11.72	25.57	0.40
	X_4	51.37	−1.99	2.15	0.40	13.83

Pooled Covariance Matrix

21.112	0.038	0.078	2.010
0.038	23.486	5.200	2.844
0.078	5.200	24.180	1.134
2.010	2.844	1.134	10.154

Note: X_1 = maximum breadth, X_2 = basibregmatic height, X_3 = basial-veolar length, X_4 = nasal height.

This only has meaning in comparison with the distances between the other pairs of samples. Calculating these as well provides the distances shown in the top part of Table 5.4.

It may be recalled from Example 4.3 that the mean values change significantly from sample to sample. The Penrose distances show that the changes are cumulative over time, with the samples that are closest in time being relatively similar, whereas the samples that are far apart in time are more different.

Turning next to Mahalanobis distances, these can be calculated from Equation 5.3, with the population covariance matrix V estimated by the pooled sample covariance matrix **C**. The matrix **C** is provided in Table 5.3, and the inverse is

$$
\mathbf{C}^{-1} = \begin{bmatrix}
0.0483 & 0.0011 & 0.0001 & -0.0001 \\
0.0011 & 0.0461 & -0.0094 & -0.0121 \\
0.0001 & -0.0094 & 0.0435 & -0.0022 \\
-0.0099 & -0.0121 & -0.0022 & 0.1041
\end{bmatrix}
$$

Using this inverse and the sample means, the Mahalanobis distance from sample 1 to sample 2 is found to be

$$
\begin{aligned}
D_{12}^2 =\ & (131.37-132.37)0.0483(131.37-132.37) \\
& +(131.37-132.37)0.0011(133.60-132.70) \\
& +\ldots-(50.53-50.23)0.0022(99.17-99.07) \\
& +(50.53-50.23)0.1041(50.53-50.23) \\
=\ & 0.091
\end{aligned}
$$

Calculating the other distances between samples in the same way provides the distance matrix shown in the lower part of Table 5.4.

A comparison between these distances and the Penrose distances shows a very good agreement. The Mahalanobis distances are three to four times as great as the Penrose distances. However, the relative distances between samples are almost the same for both measures. For example, the Penrose measure suggests that the distance from the early predynastic sample to the

Table 5.4 Penrose and Mahalanobis Distances between Pairs of Samples of Egyptian Skulls

	Early Predynastic	Late Predynastic	12–13th Dynasties	Ptolemaic	Roman
Penrose Distances					
Early predynastic	—				
Late predynastic	0.023	—			
12–13th dynasties	0.216	0.163	—		
Ptolemaic	0.493	0.404	0.108	—	
Roman	0.736	0.583	0.244	0.066	—
Mahalanobis Distances					
Early predynastic	—				
Late predynastic	0.091	—			
12–13th dynasties	0.903	0.729	—		
Ptolemaic	1.881	1.594	0.443	—	
Roman	2.697	2.176	0.911	0.219	—

Roman sample is $0.736/0.023 = 32.0$ times as great as the distance from the early predynastic to the late predynastic sample. The corresponding ratio for the Mahalanobis measure is $2.697/0.091 = 29.6$.

5.4 Distances based on proportions

A particular situation that sometimes occurs is that the variables being used to measure the distance between populations or samples are proportions that add to one. For example, the animals of a certain species might be classified into K genetic classes. One colony might then have proportions p_1 of class 1, p_2 of class 2, up to p_K of class K, while a second colony has proportions q_1 of class 1, q_2 of class 2, up to q_K of class K. The question then arises of how different is the extent of the genetic difference between the two colonies.

Various indices of distance have been proposed with this type of proportion data. For example,

$$d_1 = \sum_{i=1}^{K} |p_i - q_i|/2 \tag{5.5}$$

which is half of the sum of absolute proportion differences, is one possibility. This index takes the value one when there is no overlap of classes and the value zero when $p_i = q_i$ for all i. Another possibility is

$$d_2 = 1 - \sum_{i=1}^{K} p_i q_i \Big/ \left\{ \sum_{i=1}^{K} p_i^2 \sum q_i^2 \right\}^{1/2} \tag{5.6}$$

which again varies from one (no overlap) to zero (equal proportions).

Because d_1 and d_2 vary from zero to one, it follows that $1 - d_1$ and $1 - d_2$ are measures of the similarity between the cases being compared. In fact, it is in terms of similarities that the indices are often used. For example,

$$s_1 = 1 - d_1 = \sum |p_i - q_i|/2$$

is often used as a measure of the niche overlap between two species, where p_i is the fraction of the resources used by species 1 that are of type i and q_i is the fraction of the resources used by species 2 that are of type i. Then $s_1 = 0$ indicates that the two species use completely different resources, and $s_1 = 1$ indicates that the two species use exactly the same resources.

A similarity measure can also be constructed from any distance measure D that varies from zero to infinity. Taking $S = 1/D$ gives a similarity that ranges from infinity for two items that are no distance apart, to zero for two objects that are infinitely far apart. Alternatively, $1/(1 + D)$ ranges from 1 when $D = 0$, to 0 when D is infinite.

Table 5.5 Presences and Absences of Two Species at Ten Sites

Site	1	2	3	4	5	6	7	8	9	10
Species 1	0	0	1	1	1	0	1	1	1	0
Species 2	1	1	1	1	0	0	0	0	1	1

Note: 1 = presence, 0 = absence.

Table 5.6 Presence and Absence Data Obtained for Two Species at n Sites

	Species 2		
	Present	Absent	Total
Species 1			
Present	a	b	a + b
Absent	c	d	c + d
Total	a + c	b + d	n

5.5   Presence–absence data

Another common situation is where the similarity or distance between two items must be based on a list of their presences and absences. For example, there might be interest in the similarity between two plant species in terms of their distributions at ten sites. The data might then take the form shown in Table 5.5. Such data are often summarized as shown in Table 5.6 as counts of the number of times that both species are present (a), only one species is present (b and c), or both species are absent (d). Thus for the data in Table 5.5, $a = 3$, $b = 3$, $c = 3$, and $d = 1$.

In this situation, some of the commonly used similarity measures are:

Simple matching index $\quad = (a+d)/n$

Ochiai index $\quad = a/\{(a+b)(a+c)\}^{1/2}$

Dice–Sorensen index $\quad = 2a/(2a+b+c)$

and

Jaccard index $\quad = a/(a+b+c)$

These indices all vary from zero (no similarity) to one (complete similarity), so that complementary distance measures can be calculated by subtracting the similarity indices from one. These and other indices are reviewed by Gower and Legendre (1986), while Jackson et al. (1989) compare the results of using different indices with various multivariate analyses of the presences and absences of 25 fish species in 52 lakes.

There has been some debate about whether the number of joint absences (d) should be used in the calculation because of the danger of concluding that two species are similar simply because they are both absent from many sites. This is certainly a valid point in many situations, and it suggests that the simple matching index should be used with caution.

5.6   The Mantel randomization test

A useful test for comparing two distance or similarity matrices was introduced by Mantel (1967) as a solution to the problem of detecting space and time clustering of diseases; i.e., whether cases of a disease that occur close in space also tend to be close in time.

To understand the nature of the procedure, the following simple example should be helpful. Suppose that four objects are being studied, and that two sets of variables have been measured for each of these. The first set of variables can then be used to construct a 4×4 matrix where the entry in the ith row and jth column is a measure of the distance between object i and object j based on these variables. The distance matrix might, for example, be

$$\mathbf{M} = \begin{bmatrix} m_{11} & m_{12} & m_{13} & m_{14} \\ m_{21} & m_{22} & m_{23} & m_{24} \\ m_{31} & m_{32} & m_{33} & m_{34} \\ m_{41} & m_{42} & m_{43} & m_{44} \end{bmatrix} = \begin{bmatrix} 0.0 & 1.0 & 1.4 & 0.9 \\ 1.0 & 0.0 & 1.1 & 1.6 \\ 1.4 & 1.1 & 0.0 & 0.7 \\ 0.9 & 1.6 & 0.7 & 0.0 \end{bmatrix}$$

This is a symmetric matrix because, for example, the distance from object 2 to object 3 must be the same as the distance from object 3 to object 2 (1.1 units). Diagonal elements are zero because these represent distances from objects to themselves.

The second set of variables can also be used to construct a matrix of distances between the objects. For the example, this will be taken as

$$\mathbf{E} = \begin{bmatrix} e_{11} & e_{12} & e_{13} & e_{14} \\ e_{21} & e_{22} & e_{23} & e_{24} \\ e_{31} & e_{32} & e_{33} & e_{34} \\ e_{41} & e_{42} & e_{43} & e_{44} \end{bmatrix} = \begin{bmatrix} 0.0 & 0.5 & 0.8 & 0.6 \\ 0.5 & 0.0 & 0.5 & 0.9 \\ 0.8 & 0.5 & 0.0 & 0.4 \\ 0.6 & 0.9 & 0.4 & 0.0 \end{bmatrix}$$

Like $\mathbf{M}$, this is symmetric, with zeros down the diagonal.

Mantel's test is concerned with assessing whether the elements in $\mathbf{M}$ and $\mathbf{E}$ show some significant correlation. The test statistic that is used is sometimes the correlation between the corresponding elements of the two matrices (matching m_{11} with e_{11}, m_{12} with e_{12}, etc.), or the simpler sum of the products of these matched elements. For the general case of n × n matrices, the latter statistic is then

$$Z = \sum_{i=2}^{n} \sum_{j=1}^{i-1} m_{ij} e_{ij} \qquad (5.7)$$

This statistic is calculated and compared with the distribution of Z that is obtained by taking the objects in a random order for one of the matrices, which is why it is called a randomization test.

For the randomization test, the matrix **M** can be left as it is. A random order can then be chosen for the objects for matrix **E**. For example, suppose that a random ordering of objects turns out to be 3,2,4,1. This then gives a randomized **E** matrix of

$$\mathbf{E} = \begin{bmatrix} 0.0 & 0.5 & 0.4 & 0.8 \\ 0.5 & 0.0 & 0.9 & 0.5 \\ 0.4 & 0.9 & 0.0 & 0.6 \\ 0.8 & 0.5 & 0.6 & 0.0 \end{bmatrix}$$

The entry in row 1 and column 2 is 0.5, the distance between objects 3 and 2; the entry in row 1 and column 3 is 0.4, the distance between objects 3 and 4; and so on. A Z value can be calculated using **M** and $\mathbf{E_R}$. Repeating this procedure using different random orders of the objects for $\mathbf{E_R}$ generates the randomized distribution of Z. A check can then be made to see whether the observed Z value is a typical value from this distribution.

The basic idea here is that if the two measures of distance are quite unrelated, then the matrix **E** will be just like one of the randomly ordered matrices $\mathbf{E_R}$. Hence the observed Z will be a typical randomized Z value. On the other hand, if the two distance measures have a positive correlation, then the observed Z will tend to be larger than values given by randomization. A negative correlation between distances should not occur, but if it does, then the result will be that the observed Z value will tend to be low when compared with the randomized distribution.

With n objects, there are n! (n factorial) different possible orderings of the object numbers. There are therefore n! possible randomizations of the elements of **E**, some of which might give the same Z values. Hence, in the example with four objects, the randomized Z distribution has 4! = 24 equally likely values. It is not too difficult to calculate all of these. More realistic cases might involve, say, 15 objects, in which case the number of possible Z values is $15! \approx 1.3 \times 10^{12}$. Enumerating all of these then becomes impractical, and there are two possible approaches for carrying out the Mantel test. A large number of randomized $\mathbf{E_R}$ matrices can be generated on a computer, and the resulting distribution of Z values can be used in place of the true randomized distribution. Alternatively, the mean, E(Z), and variance, Var(Z), of the randomized distribution of Z can be calculated, and

$$g = \{Z - E(Z)\}/Var(Z)$$

can be treated as a standard normal variate. Mantel (1967) provided formulae for the mean and variance of Z in the null-hypothesis case of no correlation between the distance measures. There is, however, some doubt about the validity of the normal approximation for the test statistic g (Mielke, 1978). Given the ready availability of computers, it therefore seems best to perform randomizations rather than to rely on this approximation.

The test statistic Z of Equation 5.7 is the sum of the products of the elements in the lower diagonal parts of the matrices **M** and **E**. The only reason for using this particular statistic is that Mantel's equations for the mean and variance are available. However, if it is decided to determine significance by computer and randomizations, there is no particular reason why the test statistic should not be changed. Indeed, values of Z are not particularly informative except in comparison with the randomization mean and variance. It may therefore be more useful to take the correlation r_{ME} between the lower diagonal elements of **M** and **E** as the test statistic instead of Z. With n × n matrices, there are n(n – 1)/2 lower diagonal terms, which pair up as (m_{21}, e_{21}), (m_{31}, e_{31}), (m_{32}, e_{32}), and so on. Their correlation is calculated in the usual way, as explained in Section 2.7.

The correlation r_{ME} has the usual interpretation in terms of the relationship between the two distance measures. Thus r lies in the range from –1 to +1, with r = –1 indicating a perfect negative correlation, r = 0 indicating no correlation, and r = +1 indicating a perfect positive correlation. The significance or otherwise of the data will be the same for the test statistics Z and r because, in fact, there is a simple linear relationship between them.

Example 5.3   *More on distances between samples of Egyptian skulls*

Returning to the Egyptian skull data, we can ask the question of whether the distances given in Table 5.4, based upon four skull measurements, are significantly related to the time differences between the five samples. This certainly does seem to be the case, but a definitive answer is provided by Mantel's test.

The sample times are approximately 4000 B.C. (early predynastic), 3300 B.C. (late predynastic), 1850 B.C. (12th and 13th dynasties), 200 B.C. (Ptolemaic), and A.D. 150 (Roman). Comparing Penrose's distance measures with time differences (in thousands of years) therefore provides the lower diagonal distance matrices between the samples that are shown in Table 5.7. The correlation between the elements of these matrices is 0.954. It appears, therefore, that the distances agree very well.

Table 5.7 Penrose Distances Based on Skull Measurements and Time Differences (thousands of years) for Five Samples of Egyptian Skulls

Penrose Distances					Time Distances				
—					—				
0.023	—				0.70	—			
0.216	0.163	—			2.15	1.45	—		
0.493	0.404	0.108	—		3.80	3.10	1.65	—	
0.736	0.583	0.244	0.066	—	4.15	3.45	2.00	0.35	—

There are 5! = 120 possible ways to reorder the five samples for one of the two matrices, and, consequently, there are 120 elements in the randomization distribution for the correlation. Of these, one is the observed correlation of 0.954 and another is a larger correlation. It follows that the observed correlation is significantly high at the (2/120)100% =1.7% level, and there is evidence of a relationship between the two distance matrices. A one-sided test is appropriate because there is no reason why the samples of skulls should become more similar as they get further apart in time.

The matrix correlation between Mahalanobis distances and time distance is 0.964. This is also significantly large at the 1.7% level when compared with the randomization distribution.

5.7 Computer programs

The calculation of distance and similarity measures is the first step in the analysis of multivariate data using cluster analysis and ordination methods. For this reason, the calculation of these measures is often easiest to do using computer programs that are designed for these methods, or the clustering and ordination options of more general statistical packages.

Computer programs for Mantel's test on distance and similarity matrices were hard to find at one time. Now a Web search will produce several sites such as the Rundom Projects Page (Jadwiszczak, 2003), where programs can be downloaded, some of them apparently at no cost. Also, some standard statistical packages allow this test as an option, as detailed in the Appendix.

5.8 Discussion and further reading

The use of different measures of distance and similarity is the subject of continuing debate, indicating a lack of agreement about what is the best under different circumstances. The problem is that no measure is perfect, and the conclusions from an analysis may depend to some extent on which of several reasonable measures is used. The situation depends very much on what the purpose is for calculating the distances or similarities as well as the nature of the data available.

The usefulness of the Mantel randomization method for testing for an association between two distance or similarity matrices has led to a number of proposals for methods to analyze relationships between three or more such matrices. These are reviewed by Manly (1997). At present, a major unresolved problem in this area relates to the question of how to take proper account of the effects of spatial correlation when, as is often the case, the items that distances and similarities are measured between tend to be similar when they are relatively close in space.

Recently, Peres-Neto and Jackson (2001) have suggested that the comparison between two distance matrices using a method called Procrustes analysis is better than the use of the Mantel test. Procrustes analysis was developed by Gower (1971) as a means of seeing how well two data configurations can be matched up after suitable manipulations. Peres-Neto and Jackson propose a randomization test to assess whether the matching that can be obtained with two distance matrices is significantly better than expected by chance.

5.9   Chapter summary

- Many multivariate problems can be considered in terms of distances either between pairs of observations, samples of observations, or populations of observations.
- The Euclidean distance between pairs of individual observations is one measure of distance that is often used. This can be viewed with two or three variables as the spatial distance between the individual observations when they are plotted. This concept is generalized for use with more than three variables.
- Two measures of the distance between two multivariate samples or two multivariate populations are described. These are the Penrose distance, which does not take into account the correlations between variables, and the Mahalanobis distance, which does take the correlations into account.
- Two measures of distance are described for the situation where the p variables measured on the objects being considered are proportions that add to one. These are converted to measures of similarity by subtracting them from one.
- Four indices are described for measuring the similarity between the objects being compared, on the basis of the presence and absence of a number of characteristics.
- The Mantel test is described as a means of determining whether two matrices of distances or similarities show a significant positive or negative association.
- The availability of computer programs for computing distance and similarity measures as well as for carrying out Mantel's test are discussed.
- Problems related to the decision about how to measure multivariate distances, the extension of the Mantel randomization test to situations with three or more matrices, and Procrustes analysis as an alternative to a Mantel test with two matrices are mentioned.

Exercise

Consider the data in Table 1.3.

1. Standardize the environmental variables (altitude, annual precipitation, annual maximum temperature, and annual minimum temperature) to means of zero and standard deviations of one, and calculate Euclidean distances between all pairs of colonies using Equation 5.1 to obtain an environmental distance matrix.
2. Use the Pgi gene frequencies, converted to proportions, to calculate genetic distances between the colonies using Equation 5.5.
3. Carry out a Mantel matrix randomization test to determine whether there is a significant positive relationship between the environmental and genetic distances and report your conclusions.
4. Explain why a significant positive relationship on a randomization test in a situation such as this could be the result of spatial correlations between the data for close colonies rather than from environmental effects on the genetic composition of colonies.

References

Gower, J.C. (1971), Statistical methods for comparing different multivariate analyses of the same data, in *Mathematics in the Archaeological and Historical Sciences*, Hodson, F.R., Kendall, D.G., and Tautu, P., Eds., Edinburgh University Press, Edinburgh, pp. 138–149.

Gower, J.C. and Legendre, P. (1986), Metric and non-metric properties of dissimilarity coefficients, *J. Classification*, 5, 5–48.

Higham, C.F.W., Kijngam, A., and Manly, B.F.J. (1980), Analysis of prehistoric canid remains from Thailand, *J. Archaeological Sci.*, 7, 149–165.

Jackson, D.A., Somers, K.M., and Harvey, H.H. (1989), Similarity coefficients: measures of co-occurrence and association or simply measures of co-occurrence, *Am. Naturalist*, 133, 436–453.

Jadwiszczak, P. (2003), The Rundom Projects Page; available on-line at pjadw.tripod.com/rpabout.htm.

Mahalanobis, P.C. (1948), Historic note on the D^2 statistic, *Sankhya*, 9, 237.

Manly, B.F.J. (1997), *Randomization, Bootstrap and Monte Carlo Methods in Biology*, 2nd ed., Chapman and Hall, London.

Mantel, N. (1967), The detection of disease clustering and a generalized regression approach, *Cancer Res.*, 27, 209–220.

Mielke, P.W. (1978), Classification and appropriate inferences for Mantel and Varland's nonparametric multivariate analysis technique, *Biometrics*, 34, 272–282.

Penrose, L.W. (1953), Distance, size and shape, *Ann. Eugenics*, 18, 337–343.

Peres-Neto, P.R. and Jackson, D.A. (2001), How well do multivariate data sets match? The advantages of a Procrustean superimposition approach over a Mantel test, *Oecologia*, 129, 169–178.

chapter six

Principal components analysis

6.1 Definition of principal components

The technique of principal components analysis was first described by Karl Pearson (1901). He apparently believed that this was the correct solution to some of the problems that were of interest to biometricians at that time, although he did not propose a practical method of calculation for more than two or three variables. A description of practical computing methods came much later from Hotelling (1933). Even then, the calculations were extremely daunting for more than a few variables because they had to be done by hand. It was not until electronic computers became generally available that the principal components technique achieved widespread use.

Principal components analysis is one of the simplest of the multivariate methods. The objective of the analysis is to take p variables $X_1, X_2, ..., X_p$ and find combinations of these to produce indices $Z_1, Z_2, ..., Z_p$ that are uncorrelated in order of their importance, and that describe the variation in the data. The lack of correlation means that the indices are measuring different "dimensions" of the data, and the ordering is such that $\text{Var}(Z_1) \geq \text{Var}(Z_2) \geq ... \geq \text{Var}(Z_p)$, where $\text{Var}(Z_i)$ denotes the variance of Z_i. The Z indices are then the principal components. When doing a principal components analysis, there is always the hope that the variances of most of the indices will be so low as to be negligible. In that case, most of the variation in the full data set can be adequately described by the few Z variables with variances that are not negligible, and some degree of economy is then achieved.

Principal components analysis does not always work, in the sense that a large number of original variables are reduced to a small number of transformed variables. Indeed, if the original variables are uncorrelated, then the analysis achieves nothing. The best results are obtained when the original variables are very highly correlated, positively or negatively. If that is the case, then it is quite conceivable that 20 or more original variables can be adequately represented by two or three principal components. If this

Table 6.1 Correlations between the Five Body Measurements of Female
Sparrows Calculated from the Data of Table 1.1

	X_1	X_2	X_3	X_4	X_5
X_1, total length	1.000				
X_2, alar extent	0.735	1.000			
X_3, length of beak and head	0.662	0.674	1.000		
X_4, length of humerus	0.645	0.769	0.763	1.000	
X_5, length of keel of sternum	0.605	0.529	0.526	0.607	1.000

Note: Only the lower part of the table is shown because the correlation between X_i and
       X_j is the same as the correlation between X_j and X_i.

desirable state of affairs does occur, then the important principal components
will be of some interest as measures of the underlying dimensions in the
data. It will also be of value to know that there is a good deal of redundancy
in the original variables, with most of them measuring similar things.

  Before describing the calculations involved in a principal components
analysis, it is of value to look briefly at the outcome of the analysis when it
is applied to the data in Table 1.1 on five body measurements of 49 female
sparrows. Details of the analysis are given in Example 6.1. In this case, the
five measurements are quite highly correlated, as shown in Table 6.1. This
is therefore good material for the analysis in question. It turns out that the
first principal component has a variance of 3.62, whereas the other components all have variances that are much less than this (0.53, 0.39, 0.30, and
0.16). This means that the first principal component is by far the most important of the five components for representing the variation in the measurements of the 49 birds. The first component is calculated to be

$$Z_1 = 0.45\ X_1 + 0.46\ X_2 + 0.45\ X_3 + 0.47\ X_4 + 0.40\ X_5$$

where X_1 to X_5 denote the measurements in Table 1.1 in order, after they
have been standardized to have zero means and unit standard deviations.

  Clearly, Z_1 is essentially just an average of the standardized body measurements, and it can be thought of as a simple index of size. The analysis
given in Example 6.1 therefore leads to the conclusion that most of the
differences between the 49 birds are a matter of size (rather than shape).

6.2 *Procedure for a principal components analysis*

A principal components analysis starts with data on p variables for n individuals, as indicated in Table 6.2. The first principal component is then the
linear combination of the variables X_1, X_2, ..., X_p

$$Z_1 = a_{11}X_1 + a_{12}X_2 + ... + a_{1p}X_p$$

Table 6.2 The Form of Data for a Principal Components Analysis, with Variables X_1 to X_p and Observations on n Cases

Case	X_1	X_2	...	X_p
1	x_{11}	x_{12}	...	x_{1p}
2	x_{21}	x_{22}	...	x_{2p}
.	.	.	...	.
.	.	.	...	.
.	.	.	...	.
n	x_{n1}	x_{n2}	...	x_{np}

that varies as much as possible for the individuals, subject to the condition that

$$a_{11}^2 + a_{12}^2 + \ldots + a_{1p}^2 = 1$$

Thus Var(Z_1), the variance of Z_1, is as large as possible given this constraint on the constants a_{1j}. The constraint is introduced because if this is not done, then Var(Z_1) can be increased by simply increasing any one of the a_{1j} values.

The second principal component

$$Z_2 = a_{21}X_1 + a_{22}X_2 + \ldots + a_{2p}X_p$$

is chosen so that Var(Z_2) is as large as possible subject to the constraint that

$$a_{21}^2 + a_{22}^2 + \ldots + a_{2p}^2 = 1$$

and also to the condition that Z_1 and Z_2 have zero correlation for the data. The third principal component,

$$Z_3 = a_{31}X_1 + a_{32}X_2 + \ldots + a_{3p}X_p$$

is such that Var(Z_3) is as large as possible subject to the constraint that

$$a_{31}^2 + a_{32}^2 + \ldots + a_{3p}^2 = 1$$

and also that Z_3 is uncorrelated with both Z_1 and Z_2. Further principal components are defined by continuing in the same way. If there are p variables, then there will be up to p principal components.

In order to use the results of a principal components analysis, it is not necessary to know how the equations for the principal components are derived. However, it is useful to understand the nature of the equations themselves. In fact, a principal components analysis involves finding the eigenvalues of the sample covariance matrix.

The calculation of the sample covariance matrix has been described in Section 2.7. The covariance matrix is symmetric and has the form:

$$C = \begin{bmatrix} c_{11} & c_{12} & .. & c_{1p} \\ c_{21} & c_{22} & .. & c_{2p} \\ . & . & & . \\ . & . & & . \\ c_{p1} & c_{p2} & .. & c_{pp} \end{bmatrix}$$

where the diagonal element c_{ii} is the variance of X_i, and the off-diagonal term c_{ij} is the covariance of variables X_i and X_j.

The variances of the principal components are the eigenvalues of the matrix C. There are p of these eigenvalues, some of which may be zero. Negative eigenvalues are not possible for a covariance matrix. Assuming that the eigenvalues are ordered as $\lambda_1 \geq \lambda_2 \geq \ldots \geq \lambda_p \geq 0$, then λ_i corresponds to the ith principal component

$$Z_i = a_{i1}X_1 + a_{i2}X_2 + \ldots + a_{ip}X_p$$

In particular, $Var(Z_i) = \lambda_i$, and the constants $a_{i1}, a_{i2}, \ldots, a_{ip}$ are the elements of the corresponding eigenvector, scaled so that

$$a_{i1}^2 + a_{i2}^2 + \ldots + a_{ip}^2 = 1$$

An important property of the eigenvalues is that they add up to the sum of the diagonal elements (the trace) of the matrix C. That is

$$\lambda_1 + \lambda_2 + \ldots + \lambda_p = c_{11} + c_{22} \ldots + c_{pp}$$

Because c_{ii} is the variance of X_i and λ_i is the variance of Z_i, this means that the sum of the variances of the principal components is equal to the sum of the variances of the original variables. Therefore, in a sense, the principal components account for all of the variation in the original data.

In order to avoid one or two variables having an undue influence on the principal components, it is usual to code the variables $X_1, X_2, \ldots, X_p$ to have means of zero and variances of one at the start of an analysis. The matrix C then takes the form

$$C = \begin{bmatrix} 1 & c_{12} & .. & c_{1p} \\ c_{21} & 1 & .. & c_{2p} \\ . & . & & . \\ . & . & & . \\ . & . & & . \\ c_{p1} & c_{p2} & .. & 1 \end{bmatrix}$$

where $c_{ij} = c_{ji}$, is the correlation between X_i and X_j. In other words, the principal components analysis is carried out on the correlation matrix. In that case, the sum of the diagonal terms, and hence the sum of the eigenvalues, is equal to p, the number of X variables.

The steps in a principal components analysis can now be stated:

1. Start by coding the variables X_1, X_2, ..., X_p to have zero means and unit variances. This is usual but is omitted in some cases where it is thought that the importance of variables is reflected in their variances.
2. Calculate the covariance matrix **C**. This is a correlation matrix if step 1 has been done.
3. Find the eigenvalues λ_1, λ_2, ..., λ_p and the corresponding eigenvectors a_1, a_2, ..., a_p. The coefficients of the ith principal component are then the elements of a_i, while λ_i is its variance.
4. Discard any components that account for only a small proportion of the variation in the data. For example, starting with 20 variables, it might be found that the first three components account for 90% of the total variance. On this basis, the other 17 components may reasonably be ignored.

Example 6.1 *Body measurements of female sparrows*

Some mention has already been made of what happens when a principal components analysis is carried out on the data on five body measurements of 49 female sparrows (Table 1.1). This example is now considered in more detail.

It is appropriate to begin with step 1 of the four parts of the analysis that have just been described. Standardization of the measurements ensures that they all have equal weight in the analysis. Omitting standardization would mean that the variables X_1 and X_2, which vary most over the 49 birds, would tend to dominate the principal components.

The covariance matrix for the standardized variables is the correlation matrix. This has already been given in lower triangular form in Table 6.1. The eigenvalues of this matrix are found to be 3.616, 0.532, 0.386, 0.302, and 0.165. These add to 5.000, the sum of the diagonal terms in the correlation matrix. The corresponding eigenvectors are shown in Table 6.3, standardized

Table 6.3 The Eigenvalues and Eigenvectors of the Correlation Matrix for Five
Measurements on 49 Female Sparrows

Component	Eigenvalue	Eigenvectors (coefficients for the principal components)				
		X_1	X_2	X_3	X_4	X_5
1	3.616	0.452	0.462	0.451	0.471	0.398
2	0.532	−0.051	0.300	0.325	0.185	−0.877
3	0.386	0.691	0.341	−0.455	−0.411	−0.179
4	0.302	−0.420	0.548	−0.606	0.388	0.069
5	0.165	0.374	−0.530	−0.343	0.652	−0.192

Note: The eigenvalues are the variances of the principal components. The eigenvectors give the
coefficients of the standardized X variables used to calculate the principal components.

so that the sum of the squares of the coefficients is one for each of them.
These eigenvectors then provide the coefficients of the principal components.

The eigenvalue for a principal component indicates the variance that it
accounts for out of the total variances of 5.000. Thus the first principal
component accounts for (3.616/5.000)100% = 72.3% of the total variance.
Similarly, the other principal components in order account for 10.6%, 7.7%,
6.0%, and 3.3%, respectively, of the total variance. Clearly, the first component is far more important than any of the others.

Another way of looking at the relative importance of principal components is in terms of their variance in comparison to the variance of the
original variables. After standardization, the original variables all have variances of 1.0. The first principal component therefore has a variance of 3.616
original variables. However, the second principal component has a variance
of only 0.532 of that of one of the original variables, while the other principal
components account for even less variation. This confirms the importance
of the first principal component in comparison with the others.

The first principal component is

$$Z_1 = 0.452X_1 + 0.462X_2 + 0.451X_3 + 0.471X_4 + 0.398X_5$$

where X_1 to X_5 are the standardized variables. The coefficients of the X
variables are nearly equal, and this is clearly an index of the size of the
sparrows. It seems, therefore, that about 72.3% of the variation in the data
are related to size differences among the sparrows.

The second principal component is

$$Z_2 = -0.051 X_1 + 0.300 X_2 + 0.325 X_3 + 0.185 X_4 - 0.877 X_5$$

This is mainly a contrast between variables X_2 (alar extent), X_3 (length of
beak and head), and X_4 (length of humerus) on the one hand, and variable
X_5 (length of the keel of the sternum) on the other. That is to say, Z_2 will be
high if X_2, X_3, and X_4 are high but X_5 is low. On the other hand, Z_2 will be
low if X_2, X_3, and X_4 are low but X_5 is high. Hence Z_2 represents a shape

difference between the sparrows. The low coefficient of X_1 (total length) means that the value of this variable does not affect Z_2 very much. The other principal components can be interpreted in a similar way. They therefore represent other aspects of shape differences.

The values of the principal components may be useful for further analyses. They are calculated in the obvious way from the standardized variables. Thus for the first bird, the original variable values are $x_1 = 156$, $x_2 = 245$, $x_3 = 31.6$, $x_4 = 18.5$, and $x_5 = 20.5$. These standardize to $x_1 = (156 - 157.980)/3.654 = -0.542$, $x_2 = (245 - 241.327)/5.068 = 0.725$, $x_3 = (31.6 - 31.459)/0.795 = 0.177$, $x_4 = (18.5 - 18.469)/0.564 = 0.055$, and $x_5 = (20.5 - 20.827)/0.991 = -0.330$, where in each case the variable mean for the 49 birds has been subtracted and a division has been made by the sample standard deviation for the 49 birds. The value of the first principal component for the first bird is therefore

$$Z_1 = 0.452 \times (-0.542) + 0.462 \times 0.725 + 0.451 \times 0.177 + 0.471 \times 0.055$$
$$+ 0.398 \times (-0.330)$$
$$= 0.064$$

The second principal component for the same bird is

$$Z_2 = -0.051 \times (-0.542) + 0.300 \times 0.725 + 0.325 \times 0.177 + 0.185 \times 0.055$$
$$- 0.877 \times (-0.330)$$
$$= 0.602$$

The other principal components can be calculated in a similar way.

The birds being considered were picked up after a severe storm. The first 21 of them recovered, while the other 28 died. A question of some interest is therefore whether the survivors and nonsurvivors show any differences. It has been shown in Example 4.1 that there is no evidence of any differences in mean values. However, in Example 4.2 it has been shown that the survivors seem to have been less variable than the nonsurvivors. The situation will now be considered in terms of principal components.

Table 6.4 Comparison between Survivors and Nonsurvivors in Terms of Means and Standard Deviations of Principal Components

Principal component	Mean		Standard deviation	
	Survivors	Nonsurvivors	Survivors	Nonsurvivors
1	−0.100	0.075	1.506	2.176
2	0.004	−0.003	0.684	0.776
3	−0.140	0.105	0.522	0.677
4	0.073	−0.055	0.563	0.543
5	0.023	−0.017	0.411	0.408

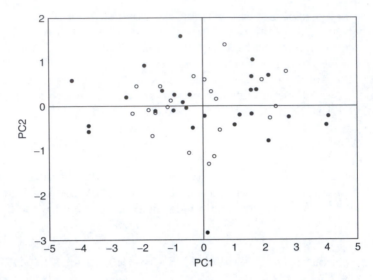

Figure 6.1 Plot of 49 female sparrows against values for the first two principal components, PC1 and PC2 (• = survivor, o = nonsurvivor).

The means and standard deviations of the five principal components are shown in Table 6.4 separately for survivors and nonsurvivors. None of the mean differences between survivors and nonsurvivors are significant from t-tests, and none of the standard deviation differences are significant on F-tests. However, Levene's test on deviations from medians (described in Chapter 4) just gives a significant difference between the variation of principal component 1 for survivors and nonsurvivors on a one-sided test at the 5% level. The assumption for the one-sided test is that, if anything, nonsurvivors were more variable than survivors. The variation is not significantly different for survivors and nonsurvivors with Levene's test on the other principal components. As principal component 1 measures overall size, it seems that stabilizing selection may have acted against very large and very small birds.

Figure 6.1 shows a plot of the values of the 49 birds for the first two principal components, which between them account for 82.9% of the variation in the data. The figure shows quite clearly how birds with extreme values for the first principal component failed to survive. Indeed, there is a suggestion that this was true for principal component 2 as well.

It is important to realize that some computer programs may give the principal components as shown with this example, but with the signs of the coefficients of the body measurements reversed. For example, Z_2 might be shown as

$$Z_2 = 0.051X_1 - 0.300X_2 - 0.325X_3 - 0.185X_4 + 0.877X_5$$

This is not a mistake. The principal component is still measuring exactly the same aspect of the data, but in the opposite direction.

Table 6.5 The Correlation Matrix for Percentages Employed in Nine Industry Groups in 30 Countries in Europe in Lower Diagonal Form, Calculated from the Data in Table 1.5

	AGR	MIN	MAN	PS	CON	SER	FIN	SPS	TC
AGR	1.000								
MIN	0.316	1.000							
MAN	-0.254	-0.672	1.000						
PS	-0.382	-0.387	0.388	1.000					
CON	-0.349	-0.129	-0.034	0.165	1.000				
SER	-0.605	-0.407	-0.033	0.155	0.473	1.000			
FIN	-0.176	-0.248	-0.274	0.094	-0.018	0.379	1.000		
SPS	-0.811	-0.316	0.050	0.238	0.072	0.388	0.166	1.000	
TC	-0.487	0.045	0.243	0.105	-0.055	-0.085	-0.391	0.475	1.000

Note: The variables are the percentages employed in AGR, agriculture, forestry, and fishing; MIN, mining and quarrying; MAN, manufacturing; PS, power and water supplies; CON, construction; SER, services; FIN, finance; SPS, social and personal services; TC, transport and communications.

Example 6.2   Employment in European countries

As a second example of a principal components analysis, consider the data in Table 1.5 on the percentages of people employed in nine industry sectors in Europe in the years from 1989 to 1995. The correlation matrix for the nine variables is shown in Table 6.5. Overall the values in this matrix are not particularly high, which indicates that several principal components will be required to account for the variation in the data.

The eigenvalues of the correlation matrix, with percentages of the total of 9.000 in parentheses, are 3.112 (34.6%), 1.809 (20.1%), 1.496 (16.6%), 1.063 (11.8%), 0.710 (7.9%), 0.311 (3.5%), 0.293 (3.3%), 0.204 (2.3%), and 0.000(0.0%). The last eigenvalue is zero because the sum of the nine variables being analyzed is 100% before standardization. The principal component corresponding to this eigenvalue has the value zero for all of the countries and hence has a zero variance. If any linear combination of the original variables in a principal components analysis is constant, then this must of necessity result in one of the eigenvalues being zero.

This example is not as straightforward as the previous one. The first principal component accounts for only about 35% of the variation in the data, and four components are needed to account for 83% of the variation. It is a matter of judgment as to how many components are important. It can be argued that only the first four should be considered because these are the ones with eigenvalues greater than one. To some extent, the choice of the number of components that are important will depend on the use that is going to be made of them. For the present example, it will be assumed that a small number of indices are required in order to present the main aspects of differences between the countries, and for simplicity only the first two components will be examined further. Between them, they account for about 55% of the variation in the original data.

The first component is

$$Z_1 = 0.51(\text{AGR}) + 0.37(\text{MIN}) - 0.25(\text{MAN}) - 0.31(\text{PS}) - 0.22(\text{CON})$$
$$- 0.38(\text{SER}) - 0.13(\text{FIN}) - 0.42(\text{SPS}) - 0.21(\text{TC})$$

where the abbreviations for variables are as defined in Table 6.5. As the analysis has been done on the correlation matrix, the variables in this equation are the original percentages after they have each been standardized to have a mean of zero and a standard deviation of one. From the coefficients of Z_1, it can be seen that it is a contrast between the numbers engaged in AGR (agriculture, forestry, and fishing) and MIN (mining and quarrying) versus the numbers engaged in other occupations.

The second component is

$$Z_2 = -0.02(\text{AGR}) + 0.00(\text{MIN}) + 0.43(\text{MAN}) + 0.11(\text{PS}) - 0.24(\text{CON})$$
$$- 0.41(\text{SER}) - 0.55(\text{FIN}) + 0.05(\text{SPS}) + 0.52(\text{TC})$$

which primarily contrasts the numbers for MAN (manufacturing) and TC (transport and communications) with the numbers in CON (construction), SER (service industries), and FIN (finance).

Figure 6.2 shows a plot of the 30 countries against their values for Z_1 and Z_2. The picture is certainly rather meaningful in terms of what is known about the countries. Most of the traditional Western democracies are grouped with slightly negative values for Z_1 and Z_2. Gibraltar and Albania stand out as having rather distinct employment patterns, while the remaining countries lie in a band ranging from the former Yugoslavia ($Z_1 = -1.2$, $Z_2 = 2.2$) to Turkey ($Z_1 = 3.2$, $Z_2 = -0.3$).

As with the previous example, it is possible that some computer programs will produce the principal components shown here, but with the signs of the coefficients of the original variables reversed. The components still measure the same aspects of the data, but with the high and low values reversed.

6.3   Computer programs

Many standard statistical packages will carry out a principal components analysis since it is one of the most common types of multivariate analysis in use. When the analysis is not mentioned as an option, it may still be possible to do the required calculations as a special type of factor analysis (as explained in Chapter 7). In that case, care will be needed to ensure that there is no confusion between the principal components and the factors, which are the principal components scaled to have unit variances.

This confusion can also occur with some programs that claim to be carrying out a principal component analysis. Instead of providing the values

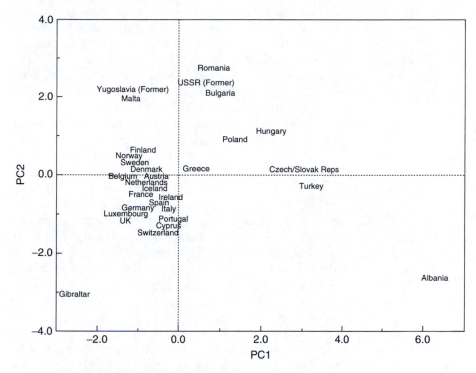

Figure 6.2 European countries plotted against the first two principal components for employment variables.

of the principal components (with variances equal to eigenvalues), they provide values of the principal components scaled to have variances of one.

6.4   Further reading

Principal components analysis is covered in almost all texts on multivariate analysis, and in greater detail by Jolliffe (2002) and Jackson (1991). Social scientists may also find the shorter monograph by Dunteman (1989) to be helpful.

6.5   Chapter summary

- Principal components analysis is described as a method for producing linear combinations of the variables $X_1, X_2, ..., X_p$ for which data are available, with the object of summarizing the main aspects of the variation in the X variables with the variation of a smaller number of these linear combinations. The linear combinations are the principal components. They take the form $Z = a_1X_1 + a_2X_2 + ... + a_pX_p$, with the constraint that $a_1^2 + a_2^2 + ... + a_p^2 = 1$.

- The first linear combination is the first principal component. This has the property of having the largest possible variance. The second principal component has the property of having the largest possible variance while being uncorrelated with the first component. The other principal components are defined similarly, with the ith principal component having the largest possible variance given that it is uncorrelated with the first i – 1 principal components.
- The principal components are calculated by finding the eigenvalues and eigenvectors of the sample covariance matrix for the X variables, usually after the X variables have been standardized to have means of zero and variances of one, so that the covariance matrix is also the correlation matrix for the X variables. If the eigenvalues are in order from the largest to the smallest, then the ith eigenvalue is the variance of the ith principal component, and the corresponding eigenvector gives the coefficients of the X variables for this principal component.
- If the analysis is carried out using the correlation matrix, then the sum of the eigenvalues is equal to p, the number of the X variables.
- For further analysis, it is usual to use only the first few principal components, providing that the sum of their variances is a high percentage (e.g., 80% or more) of the sum of the variances for all p components. Alternatively, if the analysis is carried out on the correlation matrix, then the principal components with variances greater than one may be used because these have variances that are greater than the variances of the individual standardized X variables (i.e., they account for more variation than any of the original X variables).
- The Bumpus data on five body measurements of female sparrows that survived or died as a result of a storm (Table 1.1) are analyzed. The first two principal components account for about 83% of the variation in the data. The first principal component represents the size of the sparrows, and the second principal component represents an aspect of the shape of the birds. There is some evidence that there was stabilizing selection, with moderately sized birds tending to survive better than very large or very small birds.
- The data on the percentages employed in nine different industries in Europe (Table 1.5) are also analyzed. In this case, the initial X variables are not highly correlated, and four principal components are needed to account for more than 80% of the variation in the original data. Nevertheless, a plot of the countries against their values for the first two principal components seems meaningful. For example, the traditional western democracies are mostly grouped together in the plot.

Exercises

Exercise 1

Table 6.6 shows six measurements on each of 25 pottery goblets excavated from prehistoric sites in Thailand, with Figure 6.3 illustrating the typical shape and the nature of the measurements. The main question of interest for these data concerns similarities and differences between the goblets, with obvious questions being:

> Is it possible to display the data graphically to show how the goblets are related, and if so, are there any obvious groupings of similar goblets?

Are there any goblets that are particularly unusual?

Carry out a principal components analysis and see whether the values of the principal components help to answer these questions.

Table 6.6 Measurements Taken on 25 Prehistoric Goblets from Thailand (cm)

Goblet	X_1	X_2	X_3	X_4	X_5	X_6
1	13	21	23	14	7	8
2	14	14	24	19	5	9
3	19	23	24	20	6	12
4	17	18	16	16	11	8
5	19	20	16	16	10	7
6	12	20	24	17	6	9
7	12	19	22	16	6	10
8	12	22	25	15	7	7
9	11	15	17	11	6	5
10	11	13	14	11	7	4
11	12	20	25	18	5	12
12	13	21	23	15	9	8
13	12	15	19	12	5	6
14	13	22	26	17	7	10
15	14	22	26	15	7	9
16	14	19	20	17	5	10
17	15	16	15	15	9	7
18	19	21	20	16	9	10
19	12	20	26	16	7	10
20	17	20	27	18	6	14
21	13	20	27	17	6	9
22	9	9	10	7	4	3
23	8	8	7	5	2	2
24	9	9	8	4	2	2
25	12	19	27	18	5	12

Note: The variables are defined in Figure 6.3. The data were kindly provided by Professor C.F.W. Higham of the University of Otago, New Zealand.

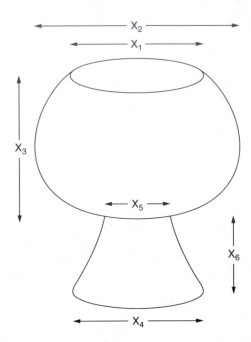

Figure 6.3 Measurements made on pottery goblets from Thailand.

One point that needs consideration with this exercise is the extent to which differences between goblets are due to shape differences rather than size differences. It may well be considered that two goblets that are almost the same shape but have very different sizes are really similar. The problem of separating size and shape differences has generated a considerable scientific literature that will not be considered here. However, it can be noted that one way to remove the effects of size involves dividing the measurements for a goblet by the total height of the body of the goblet. Alternatively, the measurements on a goblet can be expressed as a proportion of the sum of all measurements on that goblet. These types of standardization of variables will clearly ensure that the data values are similar for two goblets with the same shape but different sizes.

Exercise 2

Table 6.7 shows estimates of the average protein consumption from different food sources for the inhabitants of 25 European countries as published by Weber (1973). Use principal components analysis to investigate the relationships between the countries on the basis of these variables.

Table 6.7 Protein Consumption (g per person per day) in 25 European Countries

Country	Red Meat	White Meat	Eggs	Milk	Fish	Cereals	Starchy Foods	Pulses, Nuts, and Oilseeds	Fruit and Vegetables	Total
Albania	10	1	1	9	0.0	42	1	6	2	72
Austria	9	14	4	20	2.0	28	4	1	4	86
Belgium	14	9	4	18	5.0	27	6	2	4	89
Bulgaria	8	6	2	8	1.0	57	1	4	4	91
Czechoslovakia	10	11	3	13	2.0	34	5	1	4	83
Denmark	11	11	4	25	10.0	22	5	1	2	91
E. Germany	8	12	4	11	5.0	25	7	1	4	77
Finland	10	5	3	34	6.0	26	5	1	1	91
France	18	10	3	20	6.0	28	5	2	7	99
Greece	10	3	3	18	6.0	42	2	8	7	99
Hungary	5	12	3	10	0.0	40	4	5	4	83
Ireland	14	10	5	26	2.0	24	6	2	3	92
Italy	9	5	3	14	3.0	37	2	4	7	84
Netherlands	10	14	4	23	3.0	22	4	2	4	86
Norway	9	5	3	23	10.0	23	5	2	3	83
Poland	7	10	3	19	3.0	36	6	2	7	93
Portugal	6	4	1	5	14.0	27	6	5	8	76
Romania	6	6	2	11	1.0	50	3	5	3	87
Spain	7	3	3	9	7.0	29	6	6	7	77
Sweden	10	8	4	25	8.0	20	4	1	2	82
Switzerland	13	10	3	24	2.0	26	3	2	5	88
U.K.	17	6	5	21	4.0	24	5	3	3	88
USSR	9	5	2	17	3.0	44	6	3	3	92
W. Germany	11	13	4	19	3.0	19	5	2	4	80
Yugoslavia	4	5	1	10	1.0	56	3	6	3	89

Source: Weber, A. (1973), Agrarpolitik im Spannungsfeld der Internationalen Ernährungspolitik, Institut für Agrapolitik und Marktlehre, Kiel, Germany.

References

Dunteman, G.H. (1989), *Principal Components Analysis*, Sage Publications, Newbury Park, CA.

Hotelling, H. (1933), Analysis of a complex of statistical variables into principal components, *J. Educational Psychol.*, 24, 417–441; 498–520.

Jackson, J.E. (1991), *A User's Guide to Principal Components*, Wiley, New York.

Jolliffe, I.T. (2002), *Principal Component Analysis*, 2nd ed., Springer–Verlag, New York.

Pearson, K. (1901), On lines and planes of closest fit to a system of points in space, *Philos. Mag.*, 2, 557–572.

Weber, A. (1973), Agrarpolitik im Spannungsfeld der Internationalen Ernährungspolitik, Institut für Agrapolitik und Marktlehre, Kiel, Germany.

chapter seven

Factor analysis

7.1 The factor analysis model

Factor analysis has aims that are similar to those of principal components analysis. The basic idea is that it may be possible to describe a set of p variables $X_1, X_2, \ldots, X_p$ in terms of a smaller number of indices or factors, and in the process get a better understanding of the relationship between these variables. There is, however, one important difference. Principal components analysis is not based on any particular statistical model, whereas factor analysis is based on a model.

The early development of factor analysis is the result of work by Charles Spearman. While studying the correlations between students' test scores of various types, he noted that many observed correlations could be accounted for by a simple model (Spearman, 1904). For example, in one case, he obtained the matrix of correlations shown in Table 7.1 for boys in a preparatory school for their scores on tests in classics, French, English, mathematics, discrimination of pitch, and music. He noted that this matrix had the interesting property that any two rows are almost proportional if the diagonals are ignored. Thus for classics and English rows in Table 7.1, there are ratios:

Table 7.1 Correlations between Test Scores for Boys in a Preparatory School

	Classics	French	English	Mathematics	Discrimination of Pitch	Music
Classics	1.00	0.83	0.78	0.70	0.66	0.63
French	0.83	1.00	0.67	0.67	0.65	0.57
English	0.78	0.67	1.00	0.64	0.54	0.51
Mathematics	0.70	0.67	0.64	1.00	0.45	0.51
Discrimination of pitch	0.66	0.65	0.54	0.45	1.00	0.40
Music	–0.63	0.57	0.51	0.51	0.4.0	1.00

Source: From Spearman, C. (1904), *Am. J. Psychol.*, 15, 201–293.

$$\frac{0.83}{0.67} \approx \frac{0.70}{0.64} \approx \frac{0.66}{0.54} \approx \frac{0.63}{0.51} \approx 1.2$$

Based on this observation, Spearman suggested that the six test scores are described by the equation

$$X_i = a_i F + e_i$$

where X_i is the ith score after it has been standardized to have a mean of zero and a standard deviation of one for all of the boys. Here a_i is a constant; F is a "factor" value, which has mean of zero and standard deviation of one for all of the boys; and e_i is the part of X_i that is specific to the ith test only. Spearman showed that a constant ratio between the rows of a correlation matrix follows as a consequence of these assumptions, and that therefore this is a plausible model for the data.

Apart from the constant correlation ratios, it also follows that the variance of X_i is given by

$$\begin{aligned}
\mathrm{Var}(X_i) &= \mathrm{Var}(a_i F + e_i) \\
&= \mathrm{Var}(a_i F) + \mathrm{Var}(e_i) \\
&= a_i^2 \,\mathrm{Var}(F) + \mathrm{Var}(e_i) \\
&= a_i^2 + \mathrm{Var}(e_i)
\end{aligned}$$

because a_i is a constant, F and e_i are assumed to be independent, and the variance of F is assumed to be unity. Also, because $\mathrm{Var}(X_i) = 1$,

$$1 = a_i^2 + \mathrm{Var}(e_i)$$

Hence the constant a_i, which is called the factor loading, is such that its square is the proportion of the variance of X_i that is accounted for by the factor.

On the basis of his work, Spearman formulated his two-factor theory of mental tests. According to this theory, each test result is made up of two parts, one that is common to all the tests (general intelligence), and another that is specific to the test in question. Later this theory was modified to allow each test result to consist of a part due to several common factors plus a part specific to the test. This gives the general factor analysis model, which states that

$$X_i = a_{i1}F_1 + a_{i2}F_2 + \dots + a_{im}F_m + e_i$$

where X_i is the ith test score with mean zero and unit variance; a_{i1} to a_{im} are the factor loadings for the ith test; F_1 to F_m are m uncorrelated common factors, each with mean zero and unit variance; and e_i is a factor specific only to the ith test that is uncorrelated with any of the common factors and has zero mean.

With this model,

$$\begin{aligned} \text{Var}(X_i) = 1 &= a_{i1}^2 \, \text{Var}(F_1) + a_{i2}^2 \, \text{Var}(F_2) + \ldots + a_{im}^2 \, \text{Var}(F_m) + \text{Var}(e_i) \\ &= a_{i1}^2 + a_{i2}^2 + \ldots + a_{im}^2 + \text{Var}(e_i) \end{aligned}$$

where $a_{i1}^2 + a_{i2}^2 + \ldots + a_{im}^2$ is called the communality of X_i (the part of its variance that is related to the common factors), and $\text{Var}(e_i)$ is called the specificity of X_i (the part of its variance that is unrelated to the common factors). It can also be shown that the correlation between X_i and X_j is

$$r_{ij} = a_{i1} \, a_{j1} + a_{12} \, a_{j2} + \ldots + a_{im} \, a_{jm}$$

Hence two test scores can only be highly correlated if they have high loadings on the same factors. Furthermore, because the communality cannot exceed one, it must be the case that $-1 \le a_{ij} \le +1$.

7.2   Procedure for a factor analysis

The data for a factor analysis have the same form as for a principal components analysis. That is, there are p variables with values for these for n individuals, as shown in Table 6.2.

There are three stages to a factor analysis. To begin with, provisional factor loadings a_{ij} are determined. One approach starts with a principal components analysis and neglects the principal components after the first m, which are then taken to be the m factors. The factors found in this way are uncorrelated with each other, and are also uncorrelated with the specific factors. However, the specific factors are not uncorrelated with each other, which means that one of the assumptions of the factor analysis model does not hold. This may not matter much, providing that the communalities are high.

Whatever way the provisional factor loadings are determined, it is possible to show that they are not unique. If F_1, F_2, ..., F_m are the provisional factors, then linear combinations of these of the form

$$\begin{aligned} F_1^* &= d_{11}F_1 + d_{12}F_2 + \ldots + d_{1m}F_m \\ F_2^* &= d_{21}F_1 + d_{22}F_2 + \ldots + d_{2m}F_m \\ &\quad \cdot \\ &\quad \cdot \\ &\quad \cdot \\ F_m^* &= d_{m1}F_1 + d_{m2}F_2 + \ldots + d_{mm}F_m \end{aligned}$$

can be constructed that are uncorrelated and explain the data just as well as the provisional factors. Indeed, there are an infinite number of alternative solutions for the factor analysis model. This leads to the second stage in the analysis, which is called factor rotation. At this stage, the provisional factors are transformed in order to find new factors that are easier to interpret. To rotate or to transform in this context means essentially to choose the d_{ij} values in the above equations.

The last stage of an analysis involves calculating the factor scores. These are the values of the rotated factors $F_1^*, F_2^*, ..., F_m^*$ for each of the n individuals for which data are available.

Generally, the number of factors (m) is up to the factor analyst, although it may sometimes be suggested by the nature of the data. When a principal components analysis is used to find a provisional solution, a rough rule of thumb involves choosing m to be the number of eigenvalues greater than unity for the correlation matrix of the test scores. The logic here is the same as was explained in the previous chapter on principal components analysis. A factor associated with an eigenvalue of less than unity accounts for less variation in the data than one of the original test scores. In general, increasing m will increase the communalities of variables. However, communalities are not changed by factor rotation.

Factor rotation can be orthogonal or oblique. With orthogonal rotation, the new factors are uncorrelated, like the provisional factors. With oblique rotation, the new factors are correlated. Whichever type of rotation is used, it is desirable that the factor loadings for the new factors should be either close to zero or very different from zero. A near zero a_{ij} means that X_i is not strongly related to the factor F_j. A large positive or negative value of a_{ij} means that X_i is determined by F_j to a large extent. If each test score is strongly related to some factors, but not related to the others, then this makes the factors easier to identify than would otherwise be the case.

One method of orthogonal factor rotation that is often used is called varimax rotation. This is based on the assumption that the interpretability of factor j can be measured by the variance of the squares of its factor loadings, i.e., the variance of $a_{1j}^2, a_{2j}^2, ..., a_{mj}^2$. If this variance is large, then the a_{ij} values tend to be either close to zero or close to unity. Varimax rotation therefore maximizes the sum of these variances for all of the factors. Kaiser first suggested this approach. Later, he modified it slightly by normalizing the factor loadings before maximizing the variances of their squares, because this appears to give improved results (Kaiser, 1958). Varimax rotation can therefore be carried out with or without Kaiser normalization. Numerous other methods of orthogonal rotation have been proposed. However, varimax rotation seems to be a good standard approach.

Sometimes factor analysts are prepared to give up the idea of the factors being uncorrelated in order to make the factor loadings as simple as possible. An oblique rotation may then give a better solution than an orthogonal one. Again, there are numerous methods available to do the oblique rotation.

A method for calculating the factor scores for individuals based on principal components is described in the next section. There are other methods available, so the one chosen for use will depend on the computer package being used for an analysis.

7.3   Principal components factor analysis

It has been noted above that one way to do a factor analysis is to begin with a principal components analysis and use the first few principal components as unrotated factors. This has the virtue of simplicity, although because the specific factors $e_1, e_2, ..., e_p$, are correlated, the factor analysis model is not quite correct. Sometimes factor analysts do a principal components factor analysis first and then try other approaches afterward.

The method for finding the unrotated factors is as follows. With p variables, there will be the same number of principal components. These are linear combinations of the original variables

$$Z_1 = b_{11}X_1 + b_{12}X_2 + ... + b_{1p}X_p$$
$$Z_2 = b_{21}X_1 + b_{22}X_2 + .. + b_{2p}X_p$$
$$\cdot$$
$$\cdot \qquad\qquad\qquad\qquad\qquad\qquad (7.1)$$
$$\cdot$$
$$Z_p = b_{p1}X_1 + b_{p2}X_2 + ... + b_{pp}X_p$$

where the b_{ij} values are given by the eigenvectors of the correlation matrix. This transformation from X values to Z values is orthogonal, so that the inverse relationship is simply

$$X_1 = b_{11}Z_1 + b_{21}Z_2 + ... + b_{p1}Z_p$$
$$X_2 = b_{12}Z_1 + b_{22}Z_2 + .. + b_{p2}Z_p$$
$$\cdot$$
$$\cdot$$
$$\cdot$$
$$X_p = b_{1p}Z_1 + b_{2p}Z_2 + ... + b_{pp}Z_p$$

For a factor analysis, only m of the principal components are retained, so the last equations become

$$X_1 = b_{11}Z_1 + b_{21}Z_2 + ... + b_{m1}Z_m + e_1$$
$$X_2 = b_{12}Z_1 + b_{22}Z_2 + .. + b_{m2}Z_m + e_2$$
$$\cdot$$
$$\cdot$$
$$\cdot$$
$$X_p = b_{1p}Z_1 + b_{2p}Z_2 + ... + b_{mp}Z_m + e_p$$

where e_i is a linear combination of the principal components Z_{m+1} to Z_p. All that needs to be done now is to scale the principal components Z_1, Z_2, ..., Z_m to have unit variances, as required for factors. To do this, Z_i must be divided by its standard deviation, which is $\sqrt{\lambda_i}$, the square root of the corresponding eigenvalue in the correlation matrix. The equations then become

$$X_1 = \sqrt{\lambda_1}b_{11}F_1 + \sqrt{\lambda_2}b_{21}F_2 + ... + \sqrt{\lambda_m}b_{m1}F_m + e_1$$
$$X_2 = \sqrt{\lambda_1}b_{12}F_1 + \sqrt{\lambda_2}b_{22}F_2 + ... + \sqrt{\lambda_m}b_{m2}F_m + e_2$$
$$\vdots$$
$$X_p = \sqrt{\lambda_1}b_{1p}F_1 + \sqrt{\lambda_2}b_{2p}F_2 + ... + \sqrt{\lambda_m}b_{mp}F_m + e_p$$

where $F_i = Z_i/\sqrt{\lambda_i}$. The unrotated factor model is then

$$X_1 = a_{11}F_1 + a_{12}F_2 + ... + a_{1m}F_m + e_1$$
$$X_2 = a_{21}F_1 + a_{22}F_2 + ... + a_{2m}F_m + e_2$$
$$\vdots \tag{7.2}$$
$$X_p = a_{p1}F_1 + a_{p2}F_2 + ... + a_{pm}F_m + e_p$$

where $a_{ij} = \sqrt{\lambda_j}b_{ji}$.

After a varimax or other type of rotation, a new solution has the form

$$X_1 = g_{11}F_1^* + g_{12}F_2^* + ... + g_{1m}F_m^* + e_1$$
$$X_2 = g_{21}F_1^* + g_{22}F_2^* + ... + g_{2m}F_m^* + e_2$$
$$\vdots \tag{7.3}$$
$$X_p = g_{p1}F_1^* + g_{p2}F_2^* + ... + g_{pm}F_m^* + e_p$$

where F_i^* represents the new ith factor.

The values of the ith unrotated factor are just the values of the ith principal component after these have been scaled to have a variance of one. The values of the rotated factors are more complicated to obtain, but it can be shown that these are given by the matrix equation

$$\mathbf{F}^* = \mathbf{X}\mathbf{G}(\mathbf{G}'\mathbf{G})^{-1} \tag{7.4}$$

where $\mathbf{F}^*$ is an n × m matrix containing the values for the m rotated factors in its columns, with one row for each of the n original rows of data; $\mathbf{X}$ is the n × p matrix of the original data for p variables and n observations, after coding the variables X_1 to X_p to have means of zero and variances of one; and **G** is the p × m matrix of rotated factor loadings given by Equation 7.3.

7.4   Using a factor analysis program to do principal components analysis

Because many computer programs for factor analysis allow the option of using principal components as initial factors, it is possible to use the programs to do principal components analysis. All that has to be done is to extract the same number of factors as variables and not do any rotation. The factor loadings will then be as given by Equation 7.2, with m = p and $e_1 = e_2 = \ldots = e_p = 0$. The principal components are given by Equation 7.1, with $b_{ij} = a_{ji}/\lambda_i$, where λ_i is the ith eigenvalue.

Example 7.1   Employment in European countries

In Example 6.2, a principal components analysis was carried out on the data on the percentages of people employed in nine industry groups in 30 countries in Europe for the years 1989 to 1995 (Table 1.5). It is of some interest to continue the examination of these data using a factor analysis model.

The correlation matrix for the nine percentage variables is given in Table 6.5, and the eigenvalues and eigenvectors of this correlation matrix are shown in Table 7.2. There are four eigenvalues greater than unity, so the "rule of thumb" suggests that four factors should be considered. This is what will be done here.

The eigenvectors in Table 7.2 give the coefficients of the X-variables for Equation 7.1. These are changed into factor loadings for four factors, using Equation 7.2, to give the model

$$X_1 = + \underline{0.90}\ F_1 - 0.03\ F_2 - 0.34\ F_3 + 0.02\ F_4 + e_1\ (0.93)$$
$$X_2 = + \underline{0.66}\ F_1 + 0.00\ F_2 + \underline{0.63}\ F_3 + 0.12\ F_4 + e_2\ (0.85)$$
$$X_3 = - 0.43\ F_1 + \underline{0.58}\ F_2 - \underline{0.61}\ F_3 + 0.06\ F_4 + e_3\ (0.91)$$
$$X_4 = - \underline{0.56}\ F_1 + 0.15\ F_2 - 0.36\ F_3 + 0.02\ F_4 + e_4\ (0.46)$$
$$X_5 = - 0.39\ F_1 - 0.33\ F_2 + 0.09\ F_3 + \underline{0.81}\ F_4 + e_5\ (0.92)$$
$$X_6 = - \underline{0.67}\ F_1 - \underline{0.55}\ F_2 + 0.08\ F_3 + 0.17\ F_4 + e_6\ (0.79)$$
$$X_7 = - 0.23\ F_1 - \underline{0.74}\ F_2 - 0.12\ F_3 - \underline{0.50}\ F_4 + e_7\ (0.87)$$
$$X_8 = - \underline{0.76}\ F_1 + 0.07\ F_2 + 0.44\ F_3 - 0.33\ F_4 + e_8\ (0.88)$$
$$X_9 = - 0.36\ F_1 + \underline{0.69}\ F_2 + \underline{0.50}\ F_3 - 0.04\ F_4 + e_9\ (0.87)$$

Here, the values in parentheses are the communalities. For example, the communality for variable X_1 is $(0.90)^2 + (-0.03)^2 + (-0.34)^2 + (0.02)^2 = 0.93$.

Table 7.2 Eigenvalues and Eigenvectors for the European Employment Data of Table 1.5

	Eigenvectors								
Eigenvalue	X_1 AGR	X_2 MIN	X_3 MAN	X_4 PS	X_5 CON	X_6 SER	X_7 FIN	X_8 SPS	X_9 TC
3.111	0.512	0.375	−0.246	−0.315	−0.222	−0.382	−0.131	−0.428	−0.205
1.809	−0.024	0.000	0.432	0.109	−0.242	−0.408	−0.553	0.055	0.516
1.495	−0.278	0.516	−0.503	−0.292	0.071	0.064	−0.096	0.360	0.413
1.063	0.016	0.113	0.058	0.023	0.783	0.169	−0.489	−0.317	−0.042
0.705	0.025	−0.345	0.231	−0.854	−0.064	0.269	−0.133	0.046	0.023
0.311	−0.045	0.203	−0.028	0.208	−0.503	0.674	−0.399	−0.167	−0.136
0.293	0.166	−0.212	−0.238	0.065	0.014	−0.165	−0.463	0.619	−0.492
0.203	0.539	−0.447	−0.431	0.157	0.030	0.203	−0.026	−0.045	0.504
0.000	−0.582	−0.419	−0.447	−0.030	−0.129	−0.245	−0.191	−0.410	−0.061

Note: The variables are the percentages employed in nine industry groups: AGR, agriculture forestry and fishing; MIN, mining and quarrying; MAN, manufacturing; PS, power and water supplies; CON, construction; SER, services; FIN, finance; SPS, social and personal services; TC, transport and communications.

The communalities are quite high for all variables except X_4 (PS, power and water supplies). Most of the variance for the other eight variables is therefore accounted for by the four common factors.

Factor loadings that are 0.50 or more (ignoring the sign) are underlined in the above equations. These large and moderate loadings indicate how the variables are related to the factors. It can be seen that X_1 is almost entirely accounted for by factor 1 alone; X_2 is a mixture of factor 1 and factor 3; X_3 is accounted for by factor 1 and factor 2; etc. An undesirable property of this choice of factors is that five of the nine X variables are related strongly to two of the factors. This suggests that a factor rotation may provide a simpler model for the data.

A varimax rotation with Kaiser normalization was carried out. This produced the model

$$X_1 = + \underline{0.85}\ F_1 + 0.10\ F_2 + 0.27\ F_3 - 0.36\ F_4 + e_1$$

$$X_2 = + 0.11\ F_1 + 0.30\ F_2 + \underline{0.86}\ F_3 - 0.10\ F_4 + e_2$$

$$X_3 = - 0.03\ F_1 + 0.32\ F_2 - \underline{0.89}\ F_3 - 0.09\ F_4 + e_3$$

$$X_4 = - 0.19\ F_1 - 0.04\ F_2 - \underline{0.64}\ F_3 + 0.14\ F_4 + e_4$$

$$X_5 = - 0.02\ F_1 + 0.08\ F_2 - 0.04\ F_3 + \underline{0.95}\ F_4 + e_5$$

$$X_6 = - 0.35\ F_1 - 0.48\ F_2 - 0.15\ F_3 + \underline{0.65}\ F_4 + e_6$$

$$X_7 = - 0.08\ F_1 - \underline{0.93}\ F_2 + 0.00\ F_3 - 0.01\ F_4 + e_7$$

$$X_8 = - \underline{0.91}\ F_1 - 0.17\ F_2 - 0.12\ F_3 + 0.04\ F_4 + e_8$$

$$X_9 = - \underline{0.73}\ F_1 + \underline{0.57}\ F_2 - 0.03\ F_3 - 0.14\ F_4 + e_9$$

The communalities are unchanged and the factors are still uncorrelated. However, this is a slightly better solution than the previous one, as only X_9 is appreciably dependent on more than one factor.

At this stage, it is usual to try to put labels on factors. It is fair to say that this often requires a degree of inventiveness and imagination! In the present case, it is not too difficult, based on the highest loadings only.

Factor 1 has a high positive loading for X_1 (agriculture, forestry, and fishing) and high negative loadings for X_8 (social and personal services) and X_9 (transport and communications). It therefore measures the extent to which people are employed in agriculture rather than services and communications. It can be labeled "rural industries rather than social service and communication."

Factor 2 has high negative loadings for X_7 (finance) and a fairly high coefficient for X_9 (transport and communications). This can be labeled "lack of finance industries."

Factor 3 has a high positive loading for X_2 (mining and quarrying), a high negative loading for X_3 (manufacturing), and a moderately high negative loading for X_4 (power supplies). This can be labeled "mining rather than manufacturing."

Finally, factor 4 has a high positive loading for X_5 (construction) and a moderately high positive loading for X_6 (service industries). "Construction and service industries" seems to be a fair label in this case.

The **G** matrix of Equation 7.3 and Equation 7.4 is given by the factor loadings shown above. For example, $g_{11} = 0.85$ and $g_{12} = 0.10$, to two decimal places. Using these loadings and carrying out the matrix calculations shown in Equation 7.4 provides the values for the factor scores for each of the 30 countries in the original data set. These factor scores are shown in Table 7.3.

From studying the factor scores, it can be seen that the values for factor 1 emphasize the importance of rural industries rather than services and communications in Albania and Turkey. The values for factor 2 indicate that Bulgaria, Hungary, Romania, and the former USSR had few people employed in finance, but the Netherlands and Albania had large numbers employed in this area. The values for factor 3 contrast Albania and the Czech/Slovak Republics — with high levels of mining rather than manufacturing — with Romania and Yugoslavia, where the reverse is true. Finally, the values for factor 4 contrast Gibraltar, with high numbers in construction and service industries, with the Netherlands and Albania, where this is far from being the case.

It would be possible and reasonable to continue the analysis of this set of data by trying models with fewer factors and different methods of factor extraction. However, the general approach has been sufficiently described here, so the example will be left at this point.

It should be kept in mind by anyone attempting to reproduce the above analysis that different statistical packages may give the eigenvectors shown in Table 7.2, except that all of the coefficients have their signs reversed. A sign reversal may also occur through a factor rotation, so that the loadings

Table 7.3 Rotated Factor Scores for 30 European Countries

Country	Factor 1	Factor 2	Factor 3	Factor 4
Belgium	−0.97	−0.56	−0.10	−0.48
Denmark	−0.89	−0.47	−0.03	−0.67
France	−0.56	−0.78	−0.15	−0.25
Germany	0.05	−0.57	−0.47	0.58
Greece	0.48	0.19	−0.23	0.02
Ireland	0.28	−0.60	−0.36	0.03
Italy	0.25	−0.13	0.17	1.00
Luxembourg	−0.46	−0.36	0.02	0.92
Netherlands	−1.36	−1.56	−0.03	−2.09
Portugal	0.66	−0.45	−0.37	0.64
Spain	0.23	−0.11	−0.09	0.93
U.K.	−0.50	−1.14	−0.35	−0.04
Austria	0.18	0.05	−0.71	0.56
Finland	−0.78	−0.20	−0.21	−0.52
Iceland	−0.18	−0.04	−0.06	0.46
Norway	−1.36	−0.17	0.20	−0.42
Sweden	−1.20	−0.52	0.04	−0.74
Switzerland	0.12	−0.67	0.01	0.65
Albania	3.16	−1.82	1.76	−1.78
Bulgaria	0.47	1.56	−0.57	−0.65
Czech/Slovak Republics	−0.26	1.45	3.12	0.44
Hungary	−1.05	1.70	2.82	−0.15
Poland	0.97	0.71	−0.37	−0.42
Romania	1.11	1.73	−1.69	−0.81
USSR (former)	0.08	2.09	−0.11	0.14
Yugoslavia (former)	0.13	1.48	−1.70	0.17
Cyprus	0.46	−0.32	0.03	1.08
Gibraltar	−0.05	−1.05	0.08	3.26
Malta	−1.17	0.49	−0.79	−1.31
Turkey	2.15	0.07	0.15	−0.56

Note: Factor 1 is "rural industries rather than social service industries and communication"; factor 2 is "lack of finance industries"; factor 3 is "mining rather
than manufacturing"; and factor 4 is "construction industries."

for a rotated factor are the opposite of what is shown above. Sign reversals
like this just reverse the interpretation of the factor concerned. For example,
if the loadings for the rotated factor 1 were the opposite of those shown
above, then the results would be interpreted as social and personal services
and as transport and communications rather than rural industries.

7.5   Options in analyses

Computer programs for factor analysis often allow many different options,
which is likely to be rather confusing for the novice in this area. Typically
there might be four or five methods for the initial extraction of factors and
about the same number of methods for rotating these factors (including no

rotation). This then gives in the order of 20 different types of factor analysis that can be carried out, with results that will differ to some extent at least.

There is also the question of the number of factors to extract. Many packages will make an automatic choice, but this may or may not be acceptable. The possibility of trying different numbers of factors therefore increases the choices for an analysis even more.

On the whole, it is probably best to avoid using too many options when first practicing factor analysis. The use of principal components as initial factors with varimax rotation, as used in the example in this chapter, is a reasonable start with any set of data. The maximum-likelihood method for extracting factors is a good approach in principle, and this might also be tried if the option is available in the computer package being used.

7.6   The value of factor analysis

Factor analysis is something of an art, and it is certainly not as objective as many statistical methods. For this reason, some statisticians are skeptical about its value. For example, Chatfield and Collins (1980, p. 89) list six problems with factor analysis and conclude that "factor analysis should not be used in most practical situations." Similarly, Seber (1984) notes as a result of simulation studies that even if the postulated factor model is correct, the chance of recovering it using available methods is not high.

On the other hand, factor analysis is widely used to analyze data and, no doubt, will continue to be widely used in the future. The reason for this is that users find the results useful for gaining insight into the structure of multivariate data. Therefore, if it is thought of as a purely descriptive tool, with limitations that are understood, then it must take its place as one of the important multivariate methods. What should be avoided is carrying out a factor analysis on a single small sample that cannot be replicated and then assuming that the factors obtained must represent underlying variables that exist in the real world.

7.7   Computer programs

This chapter has stressed factor analysis based on using principal components as the unrotated factors, followed by varimax rotation. This method is widely available in computer programs, and is often the default option. There should therefore be little difficulty in obtaining suitable software if this approach is used. The use of alternative methods for factor extraction and rotation is likely to require one of the larger statistical packages that have many options, as discussed in Section 7.5. The calculations for Example 7.1 were carried out using the principal components option of NCSS (Hintze, 2001). This program has a separate factor analysis option that does the calculations in a different way and hence gives different results. However, the differences are quite small.

7.8　Discussion and further reading

Factor analysis is discussed in many texts on multivariate analysis, although as noted earlier, the topic is sometimes not presented enthusiastically (Chatfield and Collins, 1980; Seber, 1984). Recent texts are generally more positive. For example, Rencher (1995) discusses the validity of factor analysis at length as well as why it often fails to work. He notes that there are many sets of data where factor analysis should not be used, but others where the method is useful.

Factor analysis as discussed in this chapter is often referred to as exploratory factor analysis because it starts with no assumptions about the number of factors that exist or the nature of these factors. In this respect, it differs from what is called confirmatory factor analysis, which requires the number of factors and the factor structure to be specified in advance. In this way, confirmatory factor analysis can be used to test theories about the structure of the data.

Confirmatory factor analysis is more complicated to carry out than exploratory factor analysis. The details are described by Bernstein et al. (1988, Chapter 7) and Tabachnick and Fidell (2001). Confirmatory factor analysis is a special case of structural equation modeling, which is covered in chapter 14 of the latter book.

7.9　Chapter summary

- Factor analysis has aims that are similar to those of principal components analysis, i.e., to describe a set of p variables $X_1, X_2, ..., X_p$ in terms of a smaller number of indices or factors that underlie the X variables. However, the two analyses differ because factor analysis is based on a particular model for the data, whereas this is not the case for principal components analysis.

- The initial ideas about factor analysis were developed by Charles Spearman in terms of a two-factor theory for mental tests, with each test result made up of a factor common to all tests (general intelligence) and a factor specific to the test itself. The theory was then modified to give the general factor analysis model with m uncorrelated common factors, which says that the ith test result (standardized to have a mean of zero and a variance of one) is given by $X_i = a_{i1}F_1 + a_{i2}F_2 + ... + a_{im}F_m + e_i$, where F_j is the value for the jth common factor, a_{ij} is a factor loading, and e_i is the part of the test result that is specific to the ith test.

- The sum of squares of the factor loadings gives the communality (the part of the variance of the test score that is accounted for by the common factors), while $Var(e_i)$ is the specificity (the part of the variance of the test score that is unrelated to the common factors).

- The procedure for a factor analysis is described in terms of three stages. First, provisional factor loadings are determined. These are then usually modified by a process of factor rotation to obtain a model for the data that is easier to interpret. Finally, factor scores are calculated, which are the values of the factors for the individuals that have known X values.

- A principal component analysis can be used to find the initial factor solution, in which case a reasonable choice for the number of factors to use is the number of eigenvalues greater than one.
- Factor rotation can be orthogonal (to give uncorrelated factors) or oblique (to give correlated factors). The choice of a rotation method is discussed. Varimax rotation is often used.
- The procedure for carrying out a factor analysis, beginning with a principal components analysis, is described in detail.
- The use of a factor analysis program to do a principal components analysis is discussed.
- The data on the percentages employed in Europe in different industry groups (Table 1.5) is analyzed using four factors. After a varimax rotation, the factors are described as "rural industries rather than social services and communication," "lack of finance industries," "mining rather than manufacturing," and "construction and service industries."
- Computer programs for factor analysis often have many options. The use of principal components as initial factors with varimax rotation will be a reasonable start in the analysis of any data set. Other options can then be tried later.
- The general value of factor analysis is discussed. It is concluded that it is a useful tool for understanding the structure of data.
- The choice of a computer program for factor analysis is discussed.
- Further reading on factor analysis is suggested, including some on confirmatory factor analysis, where the structure of the factor analysis model is assumed to be known in advance.

Exercise

Using Example 7.1 as a model, carry out a factor analysis of the data in Table 6.7 on protein consumption from ten different food sources for the inhabitants of 25 European countries. Identify the important factors underlying the observed variables and examine the relationships between the countries with respect to these factors.

References

Bernstein, I.H., Garbin, C.P., and Teng, G.C. (1988), *Applied Multivariate Analysis*, Springer–Verlag, Berlin.

Chatfield, C. and Collins, A.J. (1980), *Introduction to Multivariate Analysis*, Chapman and Hall, London.

Hintze, J. (2001), *NCSS and PASS*, Number Cruncher Statistical Systems, Kaysville, UT; available on-line at www.ncss.com.

Kaiser, H.F. (1958), The varimax criterion for analytic rotation in factor analysis, *Psychometrika*, 23, 187–200.

Rencher, A.C. (1995), *Methods of Multivariate Statistics*, Wiley, New York.

Seber, G.A.F. (1984), *Multivariate Observations*, Wiley, New York.

Spearman, C. (1904), "General intelligence," objectively determined and measured, *Am. J. Psychol.*, 15, 201–293.

Tabachnick, B.G. and Fidell, L.S. (2001), *Using Multivariate Statistics*, Allyn and Bacon, Boston.

chapter eight

Discriminant function analysis

8.1 The problem of separating groups

The problem that is addressed with discriminant function analysis is the extent to which it is possible to separate two or more groups of individuals, given measurements for these individuals on several variables. For example, with the data in Table 1.1 on five body measurements of 21 surviving and 28 nonsurviving sparrows, it is interesting to consider whether it is possible to use the body measurements to separate survivors and nonsurvivors. Also, for the data shown in Table 1.2 on four dimensions of Egyptian skulls for samples from five time periods, it is reasonable to consider whether the measurements can be used to assign skulls to different time periods.

In the general case, there will be m random samples from different groups with sizes n_1, n_2, ... n_m, and values will be available for p variables X_1, X_2, ..., X_p for each sample member. Thus the data for a discriminant function analysis takes the form shown in Table 8.1. The data for a discriminant function analysis do not need to be standardized to have zero means and unit variances prior to the start of the analysis, as is usual with principal components and factor analysis. This is because the outcome of a discriminant function analysis is not affected in any important way by the scaling of individual variables.

8.2 Discrimination using Mahalanobis distances

One approach to discrimination is based on Mahalanobis distances, as defined in Section 5.3. The mean vectors for the m samples can be regarded as estimates of the true mean vectors for the groups. The Mahalanobis distances of the individual cases to the group centers can then be calculated, and each individual can be allocated to the group that it is closest to. This may or may not be the group that the individual actually came from, so the

Table 8.1 The Form of Data for a Discriminant Function Analysis with m Groups with Possibly Different Sizes and with p Variables Measured on Each Individual Case

Case	X_1	X_2	...	X_p	Group
1	x_{111}	x_{112}	...	x_{11p}	1
2	x_{211}	x_{212}	...	x_{21p}	1
.	.	.	.	.	.
.	.	.	.	.	.
.	.	.	.	.	.
n_1	x_{n_111}	x_{n_112}	...	x_{n_11p}	1
1	x_{121}	x_{122}	...	x_{12p}	2
2	x_{221}	x_{222}	...	x_{22p}	2
.	.	.	.	.	.
.	.	.	.	.	.
.	.	.	.	.	.
n_2	x_{n_221}	x_{n_222}	...	x_{n_22p}	2
.	.	.	.	.	.
.	.	.	.	.	.
.	.	.	.	.	.
1	x_{1m1}	x_{1m2}	...	x_{1mp}	m
2	x_{2m1}	x_{2m2}	...	x_{2mp}	m
.	.	.	.	.	.
.	.	.	.	.	.
.	.	.	.	.	.
n_m	x_{n_mm1}	x_{n_mm2}	...	x_{n_mmp}	m

percentage of correct allocations is an indication of how well groups can be separated using the available variables.

This procedure is more precisely defined as follows. Let

$$\bar{x}_i' = \left(\bar{x}_{1i}, \bar{x}_{2i}, \ldots, \bar{x}_{pi} \right)'$$

denote the vector of mean values for the sample from the ith group; let C_i denote the covariance matrix for the same sample; and let C denote the pooled sample covariance matrix, where these vectors and matrices are calculated as explained in Section 2.7. Then the Mahalanobis distance from an observation $x' = (x_1, x_2, \ldots, x_p)'$ to the center of group i is estimated to be

$$D_i^2 = \left(x - \bar{x}_i \right)' C^{-1} \left(x - \bar{x}_i \right)$$
$$= \sum_{r=1}^{p} \sum_{s=1}^{p} \left(x_r - \bar{x}_{ri} \right) c^{rs} \left(x_s - \bar{x}_{si} \right)$$

(8.1)

where c^{rs} is the element in the rth row and sth column of C^{-1}. The observation x is then allocated to the group for which D_i^2 has the smallest value.

Table 8.2 An Analysis of Variance on the Z Index

Source of variation	Degrees of freedom	Mean square	F-ratio
Between groups	$m - 1$	M_B	M_B/M_W
Within groups	$N - m$	M_W	—
	$N - 1$	—	—

8.3 Canonical discriminant functions

It is sometimes useful to be able to determine functions of the variables X_1, X_2, ..., X_p that in some sense separate the m groups as much as is possible. The simplest approach then involves taking a linear combination of the X variables

$$Z = a_1X_1 + a_2X_2 + \ldots + a_pX_p$$

for this purpose. Groups can be well separated using Z if the mean value of this variable changes considerably from group to group, with the values within a group being fairly constant.

One way to determine the coefficients a_1, a_2, ..., a_p in the index involves choosing these so as to maximize the F-ratio for a one-way analysis of variance. Thus if there are a total of N individuals in all the groups, an analysis of variance on Z values takes the form shown in Table 8.2. Hence, a suitable function for separating the groups can be defined as the linear combination for which the F-ratio M_B/M_W is as large as possible, as first suggested by Fisher (1936).

When this approach is used, it turns out that it may be possible to determine several linear combinations for separating groups. In general, the number available, s, is the smaller of p and $m - 1$. The linear combinations are referred to as canonical discriminant functions.

The first function,

$$Z_1 = a_{11}X_1 + a_{12}X_2 + \ldots + a_{1p}X_p$$

gives the maximum possible F-ratio for a one-way analysis of variance for the variation within and between groups. If there is more than one function, then the second one,

$$Z_2 = a_{21}X_1 + a_{22}X_2 + \ldots + a_{2p}X_p$$

gives the maximum possible F-ratio on a one-way analysis of variance subject to the condition that there is no correlation between Z_1 and Z_2 within groups. Further functions are defined in the same way. Thus the ith canonical discriminant function,

$$Z_i = a_{i1}X_1 + a_{i2}X_2 + \ldots + a_{ip}X_p$$

is the linear combination for which the F-ratio on an analysis of variance is maximized, subject to Z_i being uncorrelated with $Z_1, Z_2, ..., $ and Z_{i-1} within groups.

Finding the coefficients of the canonical discriminant functions turns out to be an eigenvalue problem. The within-sample matrix of sums of squares and cross-products, **W**, and the total sample matrix of sums of squares and cross-products matrix, **T**, are calculated as described in Section 4.7. From these, the between-groups matrix

$$\mathbf{B} = \mathbf{T} - \mathbf{W}$$

can be determined. Next, the eigenvalues and eigenvectors of the matrix $\mathbf{W}^{-1}\mathbf{B}$ have to be found. If the eigenvalues are $\lambda_1 > \lambda_2 > ... > \lambda_s$, then λ_i is the ratio of the between-group sum of squares to the within-group sum of squares for the ith linear combination, Z_i, while the elements of the corresponding eigenvector, $\mathbf{a}_i' = (a_{i1}, a_{i2}, ..., a_{ip})$ are the coefficients of the X variables for this index.

The canonical discriminant functions $Z_1, Z_2, ..., Z_s$ are linear combinations of the original variables chosen in such a way that Z_1 reflects group differences as much as possible, Z_2 captures as much as possible of the group differences not displayed by Z_1, Z_3 captures as much as possible of the group differences not displayed by Z_1 and Z_2, etc. The hope is that the first few functions are sufficient to account for almost all of the important group differences. In particular, if only the first one or two functions are needed for this purpose, then a simple graphical representation of the relationship between the various groups is possible by plotting the values of these functions for the sample individuals.

8.4   Tests of significance

Several tests of significance are useful in conjunction with a discriminant function analysis. In particular, the T^2-test of Section 4.3 can be used to test for a significant difference between the mean values for any pair of groups, while one of the tests described in Section 4.7 can be used to test for overall significant differences between the means for the m groups.

In addition, a test is sometimes proposed for testing whether the mean of the discriminant function Z_j differs significantly from group to group. This is based on the individual eigenvalues of the matrix $\mathbf{W}^{-1}\mathbf{B}$. For example, sometimes the statistic

$$\phi_j^2 = \{N - 1 - (p+m)/2\} \log_e (1 + \lambda_j)$$

is used, where N is the total number of observations in all groups. This statistic is then tested against the chi-squared distribution with p + m − 2j degrees of freedom (df), and a significantly large value is considered to provide evidence that the population mean values of Z_j vary from group to group. Alternatively, the sum $\phi_j^2 + \phi_{j+1}^2 + \dots + \phi_s^2$ is sometimes used for testing for group differences related to discriminant functions Z_j to Z_s. This is tested against the chi-squared distribution, with the df being the sum of those associated with the component terms. Other tests of a similar nature are also used.

Unfortunately, these tests are suspect to some extent because the jth discriminant function in the population may not appear as the jth discriminant function in the sample because of sampling errors. For example, the estimated first discriminant function (corresponding to the largest eigenvalue for the sample matrix $\mathbf{W}^{-1}\,\mathbf{B}$) may in reality correspond to the second discriminant function for the population being sampled. Simulations indicate that this can upset the chi-squared tests described above quite seriously. Therefore, it seems that the tests should not be relied upon to decide how many of the obtained discriminant functions represent real group differences. See Harris (1985) for an extended discussion of the difficulties surrounding these tests and alternative ways for examining the nature of group differences.

One useful type of test that is valid, at least for large samples, involves calculating the Mahalanobis distance from each of the observations to the mean vector for the group containing the observation, as discussed in Section 5.3. These distances should follow approximately chi-squared distributions with p degrees of freedom. Hence, if an observation is very significantly far from the center of its group in comparison with the chi-squared distribution, then this brings into question whether the observation really came from that group.

8.5   Assumptions

The methods discussed so far in this chapter are based on two assumptions. First, for all of the methods, the population within-group covariance matrix should be the same for all groups. Second, for tests of significance, the data should be multivariate normally distributed within groups.

In general it seems that multivariate analyses that assume normality may be upset quite badly if this assumption is not correct. This contrasts with the situation with univariate analyses such as regression and analysis of variance, which are generally quite robust to this assumption. However, a failure of one or both assumptions does not necessarily mean that a discriminant function analysis is a waste of time. For example, it may well turn out that excellent discrimination is possible on data from nonnormal distributions, although it may not then be simple to establish the statistical significance of the group differences. Furthermore, discrimination methods that do not require the assumptions of normality and equality of population covariance matrices are available, as discussed below.

Example 8.1   Comparison of samples of Egyptian skulls

This example concerns the comparison of the values for four measurements on male Egyptian skulls for five samples ranging in age from the early predynastic period (circa 4000 B.C.) to the Roman period (circa A.D. 150). The data are shown in Table 1.2, and it has already been established that the mean values differ significantly from sample to sample (Example 4.3), with the differences tending to increase with the time difference between samples (Example 5.3).

The within-sample and total sample matrices of sums of squares and cross-products are calculated as described in Section 4.7. They are found to be:

$$\mathbf{W} = \begin{bmatrix} 3061.67 & 5.33 & 11.47 & 291.30 \\ 5.33 & 3405.27 & 754.00 & 412.53 \\ 11.47 & 754.00 & 3505.97 & 164.33 \\ 291.30 & 412.53 & 164.33 & 1472.13 \end{bmatrix}$$

and

$$\mathbf{T} = \begin{bmatrix} 3563.89 & -222.81 & -615.16 & 426.73 \\ -222.81 & 3635,17 & 1046.28 & 346.47 \\ -615.16 & 1046.28 & 4309.27 & -16.40 \\ 426.73 & 346.47 & -16.40 & 1533.33 \end{bmatrix}$$

The between-sample matrix is therefore

$$\mathbf{B} = \mathbf{T} - \mathbf{W} = \begin{bmatrix} 502.83 & -228.16 & -626.63 & 135.43 \\ -228.15 & 229.91 & 292.28 & -66.07 \\ -626.63 & 292.28 & 803.30 & -180.73 \\ 135.43 & -66.07 & -180.73 & 61.30 \end{bmatrix}$$

The eigenvalues of $\mathbf{W}^{-1}\mathbf{B}$ are found to be $\lambda_1 = 0.437$, $\lambda_2 = 0.035$, $\lambda_3 = 0.015$, and $\lambda_4 = 0.002$, and the corresponding canonical discriminant functions are

$$\begin{aligned} Z_1 &= -0.0107X_1 + 0.0040X_2 + 0.0119X_3 - 0.0068X_4 \\ Z_2 &= 0.0031X_1 + 0.0168X_2 - 0.0046X_3 - 0.0022X_4 \\ Z_3 &= -0.0068X_1 + 0.0010X_2 + 0.0000X_3 + 0.0247X_4 \\ Z_4 &= 0.0126X_1 - 0.0001X_2 + 0.0112X_3 + 0.0054X_4 \end{aligned} \tag{8.2}$$

Because λ_1 is much larger than the other eigenvalues, it is apparent that most of the sample differences are described by Z_1 alone.

The X variables in Equation 8.2 are the values as shown in Table 1.2 without standardization. The nature of the variables is illustrated in Fig. 1.1, from which it can be seen that large values of Z_1 correspond to skulls that are tall but narrow, with long jaws and short nasal heights.

The Z_1 values for individual skulls are calculated in the obvious way. For example, the first skull in the early predynastic sample has $X_1 = 131$ mm, $X_2 = 138$ mm, $X_3 = 89$ mm, and $X_4 = 49$ mm. Therefore, for this skull

$$Z_1 = (-0.0107 \times 131) + (0.0040 \times 138) + (0.0119 \times 89) + (-0.0068 \times 49) = -0.124$$

The means and standard deviations found for the Z_1 values for the five samples are shown in Table 8.3. It can be seen that the mean of Z_1 has become lower over time, indicating a trend toward shorter, broader skulls with short jaws but relatively large nasal heights. This is, however, very much an average change. If the 150 skulls are allocated to the samples to which they are closest according to the Mahalanobis distance function of Equation 8.1, then only 51 of them (34%) are allocated to the samples that they really belong to (Table 8.4). Thus although this discriminant function analysis has been successful in pinpointing the changes in average skull dimensions over time, it has not produced a satisfactory method for estimating the age of individual skulls.

Table 8.3 Means and Standard Deviations for the Discriminant Function Z_1 with Five Samples of Egyptian Skulls

Sample	Mean	Standard deviation
Early predynastic	−0.029	0.097
Late predynastic	−0.043	0.071
12th and 13th dynasties	−0.099	0.075
Ptolemaic	−0.143	0.080
Roman	−0.167	0.095

Table 8.4 Results Obtained when 150 Egyptian Skulls Are Allocated to the Groups for Which They Have the Minimum Mahalanobis Distance

Source group	Number allocated to group					Total
	1	2	3	4	5	
1	12	8	4	4	2	30
2	10	8	5	4	3	30
3	4	4	15	2	5	30
4	3	3	7	5	12	30
5	2	4	4	9	11	30

Example 8.2   Discriminating between groups of European countries

The data shown in Table 1.5 on the percentages employed in nine industry groups in 30 European countries have already been examined by principal components analysis and by factor analysis (Examples 6.2 and 7.1). Here they will be considered from the point of view of the extent to which it is possible to discriminate between groups of countries on the basis of employment patterns. In particular, four natural groups existed in the period when the data were collected. These were: (1) the European Union (EU) countries of Belgium, Denmark, France, Germany, Greece, Ireland, Italy, Luxembourg, the Netherlands, Portugal, Spain, and the U.K.; (2) the European Free Trade Area (EFTA) countries of Austria, Finland, Iceland, Norway, Sweden, and Switzerland; (3) the eastern European countries of Albania, Bulgaria, the Czech/Slovak Republics, Hungary, Poland, Romania, the former USSR, and the former Yugoslavia; and (4) the other countries of Cyprus, Gibraltar, Malta, and Turkey. These four groups can be used as a basis for a discriminant function analysis. Wilks"s lambda test (Section 4.7) gives a very highly significant result ($p < 0.001$), so there is very clear evidence that, overall, these groups are meaningful.

Apart from rounding errors, the percentages in the nine industry groups add to 100% for each of the 30 countries. This means that any one of the nine percentage variables can be expressed as 100 minus the remaining variables. It is therefore necessary to omit one of the variables from the analysis in order to carry out the analysis. The last variable, the percentage employed in transport and communications, has therefore been omitted for the analysis that will now be described.

The number of canonical variables is three in this example, this being the minimum of the number of variables ($p = 8$) and the number of groups minus one ($m - 1 = 3$). These canonical variables are found to be

$$Z_1 = 0.427\,AGR + 0.295\,MIN + 0.359\,MAN + 0.339\,PS + 0.222\,CON$$
$$+\,0.688\,SER + 0.464\,FIN + 0.514\,SPS$$
$$Z_2 = 0.674\,AGR + 0.579\,MIN + 0.550\,MAN + 1.576\,PS + 0.682\,CON$$
$$+\,0.658\,SER + 0.349\,FIN + 0.682\,SPS$$
$$Z_3 = 0.732\,AGR + 0.889\,MIN + 0.873\,MAN + 0.410\,PS + 0.524\,CON$$
$$+\,0.895\,SER + 0.714\,FIN + 0.764\,SPS$$

Different computer programs are likely to output these canonical variables with all of the signs reversed for the coefficients of one or more of the variables. Also, it may be desirable to reverse the signs that are output. Indeed, with this example, the output from the computer program had negative coefficients for all of the variables with Z_1 and Z_2. The signs were therefore all reversed to make the coefficients positive. It is important to note that it is the original percentages employed that are to be used in these

Table 8.5 Correlations between the
Original Percentages in Different
Employment Groups and the Three
Canonical Variates

Group	Z_1	Z_2	Z_3
AGR	−0.50	0.37	0.09
MIN	−0.62	0.03	0.20
MAN	−0.02	−0.20	0.12
PS	0.17	0.18	−0.23
CON	0.14	0.26	−0.34
SER	0.82	−0.01	0.08
FIN	0.61	−0.36	−0.09
SPS	0.56	−0.19	−0.28
TC	−0.22	−0.47	−0.41

Note: AGR, agriculture, forestry, and fishing;
MIN, mining and quarrying; MAN, manufacturing; PS, power and water supplies; CON, construction; SER, services;
FIN, finance; SPS, social and personal services; TC, transport and communications.

equations, rather than these percentages after they have been standardized
to have zero means and unit variances.

The eigenvalues of $\mathbf{W}^{-1}\,\mathbf{B}$ corresponding to the three canonical variables
are $\lambda_1 = 5.349$, $\lambda_2 = 0.570$, and $\lambda_3 = 0.202$. The first canonical variable is
therefore clearly the most important.

Because all of the coefficients are positive for all three canonical variables,
it is difficult to interpret what exactly they mean in terms of the original
variables. It is helpful in this respect to consider instead the correlations
between the original variables and the canonical variables, as shown in
Table 8.5. This table includes the original variable TC (transport and communications) because the correlations for this variable are easily calculated
once the values of Z_1 to Z_3 are known for all of the European countries.

It can be seen that the first canonical variable has correlations above 0.5
for SER (services), FIN (finance), and SPS (social and personal services), and
a correlation of −0.5 or less for AGR (agriculture, forestry, and fisheries) and
MIN (mining). This canonical variable therefore represents service types of
industry rather than traditional industries. There are no really large positive
or negative correlations between the second canonical variate and the original variables. However, considering the largest correlations that there are,
it seems to represent agriculture and construction, with an absence of transport, communications, and financial services. Finally, the third canonical
variable also shows no large correlations but represents, if anything, an
absence of transport, communication, and construction.

Plots of the countries against their values for the canonical variables are
shown in Figure 8.1. The plot of the second canonical variable against the
first one shows a clear distinction between the eastern countries on the

left-hand side and the other groups on the left. There is no clear separation between the EU and EFTA countries, with Malta and Cyprus being in the same cluster. Turkey and Gibraltar from the "other" group of countries appear in the top at the right-hand side. It can be clearly seen how most separation occurs with the horizontal values for the first canonical variate. Based on the interpretation of the canonical variables given earlier, it appears that in the eastern countries there is an emphasis on traditional industries rather than service industries, whereas the opposite tends to be true for the other countries. Similarly, Turkey and Gibraltar stand out because of the emphasis on agriculture and construction rather than transport, communications, and financial services. For Gibraltar, there are apparently none engaged in agriculture, but a very high percentage in construction.

The plot of the third canonical variable against the first one shows no real vertical separation of the EU, EFTA, and other groups of countries, although there are some obvious patterns, like the Scandinavian countries appearing close together.

The discriminant function analysis has been successful in this example in separating the eastern countries from the others, with less success in separating the other groups. The separation is perhaps clearer than what was obtained using principal components, as shown in Figure 6.2.

8.6   Allowing for prior probabilities of group membership

Computer programs often allow many options for a discriminant function analysis. One situation is that the probability of membership is inherently different for different groups. For example, if there are two groups, it might be that it is known that most individuals fall into group 1, while very few fall into group 2. In that case, if an individual is to be allocated to a group, it makes sense to bias the allocation procedure in favor of group 1. Thus the process of allocating an individual to the group to which it has the smallest Mahalanobis distance should be modified. To allow for this, some computer programs permit prior probabilities of group membership to be taken into account in the analysis.

8.7   Stepwise discriminant function analysis

Another possible modification of the basic analysis involves carrying it out in a stepwise manner. In this case, variables are added to the discriminant functions one by one until it is found that adding extra variables does not give significantly better discrimination. There are many different criteria that can be used for deciding on which variables to include in the analysis and which to exclude.

A problem with stepwise discriminant function analysis is the bias that the procedure introduces into significance tests. Given enough variables, it is almost certain that some combination of them will produce significant discriminant functions by chance alone. If a stepwise analysis is carried out,

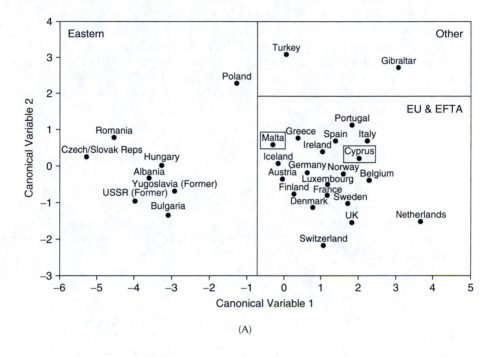

(A)

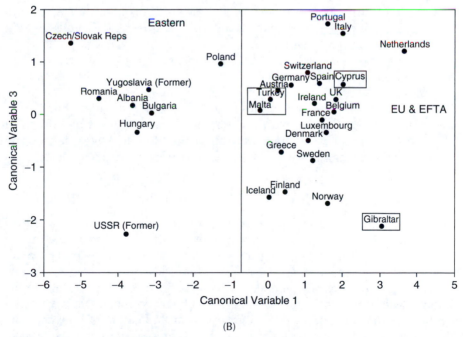

(B)

Figure 8.1 Plot of 30 European countries against their values for the first three canonical discriminant functions. Small boxes indicate countries in the other category that are not separated from the EU and EFTA groups.

then it is advisable to check its validity by rerunning it several times with a random allocation of individuals to groups to see how significant are the results obtained. For example, with the Egyptian skull data, the 150 skulls could be allocated completely at random to five groups of 30, the allocation being made a number of times, and a discriminant function analysis run on each random set of data. Some idea could then be gained of the probability of getting significant results through chance alone.

This type of randomization analysis to verify a discriminant function analysis is unnecessary for a standard non-stepwise analysis, providing that there is no reason to suspect the assumptions behind the analysis. It could, however, be informative in cases where the data are clearly not normally distributed within groups or where the within-group covariance matrix is not the same for each group. For example, Manly (1997, example 12.4) shows a situation where the results of a standard discriminant function analysis are clearly suspect by comparison with the results of a randomization analysis.

8.8   *Jackknife classification of individuals*

A moment's reflection will suggest that an allocation matrix such as that shown in Table 8.4 must tend to have a bias in favor of allocating individuals to the group that they really come from. After all, the group means are determined from the observations in that group. It is not surprising to find that an observation is closest to the center of a group where that observation helped in determining that center.

To overcome this bias, some computer programs carry out what is called a jackknife classification of observations. This involves allocating each individual to its closest group without using that individual to help determine a group center. In this way, any bias in the allocation is avoided. In practice, there is often not a great deal of difference between the straightforward classification and the jackknife classification, with the jackknife classification usually giving a slightly smaller number of correct allocations.

8.9   *Assigning of ungrouped individuals to groups*

Some computer programs allow the input of data values for a number of individuals for which the true group is not known. It is then possible to assign these individuals to the group that they are closest to, in the Mahalanobis distance sense, on the assumption that they have to come from one of the m groups that are sampled. Obviously, in these cases it will not be known whether the assignment is correct. However, the errors in the allocation of individuals from known groups is an indication of how accurate the assignment process is likely to be. For example, the results shown in Table 8.4 indicate that allocating Egyptian skulls to different time periods using skull dimensions is liable to result in many errors.

8.10   Logistic regression

A rather different approach to discrimination between two groups involves making use of logistic regression. In order to explain how this is done, the more usual use of logistic regression will first be briefly reviewed.

The general framework for logistic regression is that there are m groups to be compared, with group i consisting of n_i items, of which Y_i exhibit a positive response (a success) and $n_i - Y_i$ exhibit a negative response (a failure). The assumption is then made that the probability of a success for an item in group i is given by

$$\pi_i = \frac{\exp\left(\beta_0 + \beta_1 x_{i1} + \beta_2 x_{i2} + \ldots + \beta_p x_{ip}\right)}{1 + \exp\left(\beta_0 + \beta_1 x_{i1} + \beta_2 x_{i2} + \ldots + \beta_p x_{ip}\right)} \tag{8.3}$$

where x_{ij} is the value of some variable X_j that is the same for all items in the group. In this way, the variables X_1 to X_p are allowed to influence the probability of a success, which is assumed to be the same for all items in the group, irrespective of the successes or failures of the other items in that or any other group. Similarly, the probability of a failure is $1 - \pi_i$ for all items in the ith group. It is permissible for some or all of the group to contain just one item. Indeed, some computer programs allow for only this to be the case.

There need be no concern about arbitrarily choosing what to call a success and what to call a failure. It is easy to show that reversing these designations in the data simply results in all the β values and their estimates changing sign, and consequently changing π_i into $1 - \pi_i$.

The function that is used to relate the probability of a success to the X variables is called a logistic function. Unlike the standard multiple regression function, the logistic function forces estimated probabilities to lie within the range zero to one. It is for this reason that logistic regression is more sensible than linear regression as a means of modeling probabilities.

There are numerous computer programs available for fitting Equation 8.3 to data, i.e., for estimating the values of β_0 to β_p. They are commonly based on the principle of maximum likelihood, which means that the equations for the estimation of the β values do not have an explicit solution. As a result, the calculations involve an iterative process of improving initial approximations for the estimates until no further changes can be made. The output commonly includes the estimates of the β values and their standard errors, a chi-squared statistic that indicates the extent to which the model fits the data, and a chi-squared statistic that indicates the extent to which the model is an improvement over what is obtained by assuming that the probability of a success is unrelated to the X variables.

In the context of discrimination with two samples, there are three different types of situation that have to be considered:

1. The data consist of a single random sample taken from a population of items that is itself divided into two parts. The application of logistic regression is then straightforward, and the fitted Equation 8.3 can be used to give an estimate of the probability of an item being in one part of the population (i.e., is a success) as a function of the values that the item possesses for variables X_1 to X_p. In addition, the distribution of success probabilities for the sampled items is an estimate of the distribution of these probabilities for the full population.

2. Separate sampling is used, where a random sample of size n_1 is taken from the population of items of one type (the successes), and an independent random sample of size n_2 is taken from the population of items of the second type (the failures). Logistic regression can still be used. However, the estimated probability of a success obtained from the estimated function must be interpreted in terms of the sampling scheme and the sample sizes used.

3. Groups of items are chosen to have particular values for the variables X_1 to X_p, such that these variable values change from group to group. The number of successes in each group is then observed. In this case, the estimated logistic regression equation gives the probability of a success for an item, conditional on the values that the item possesses for X_1 to X_p. The estimated function is therefore the same as for situation 1, but the sample distribution of probabilities of a success is in no way an estimate of the distribution that would be found in the combined population of items that are successes or failures.

The following examples illustrate the differences between situations 1 and 2, which are the ones that most commonly occur. Situation 3 is really just a standard logistic regression situation and will not be considered further here.

Example 8.3   Storm survival of female sparrows (reconsidered)

The data in Table 1.1 consist of values for five morphological variables for 49 female sparrows taken in a moribund condition to Hermon Bumpus's laboratory at Brown University in Rhode Island after a severe storm in 1898. The first 21 birds recovered, and the remaining 28 died, and there is some interest in knowing whether it is possible to discriminate between these two groups on the basis of the five measurements. It has already been shown that there are no significant differences between the mean values of the variables for survivors and nonsurvivors (Example 4.1), although the nonsurvivors may have been more variable (Example 4.2). A principal components analysis has also confirmed the test results (Example 6.1).

This is a situation of type 1 if the assumption is made that the sampled birds were randomly selected from the population of female sparrows in some area close to Bumpus's laboratory. Actually, the assumption of random sampling is questionable because it is not clear exactly how the birds were collected. Nevertheless, the assumption will be made for this example.

The logistic regression option in many standard computer packages can be used to fit the model

$$\pi_i = \frac{\exp\left(\beta_0 + \beta_1 x_{i1} + \beta_2 x_{i2} + \ldots + \beta_5 x_{i5}\right)}{1 + \exp\left(\beta_0 + \beta_1 x_{i1} + \beta_2 x_{i2} + \ldots + \beta_5 x_{i5}\right)}$$

where the variables are X_1 = total length, X_2 = alar extent, X_3 = length of beak and head, X_4 = length of the humerus, and X_5 = length of the sternum (all in mm), and π_i denotes the probability of the ith bird recovering from the storm.

A chi-squared test for whether the variables account significantly for the difference between survivors and nonsurvivors gives the value 2.85 with five degrees of freedom, which is not at all significantly large when compared with chi-squared tables. There is therefore no evidence from this analysis that the survival status was related to the morphological variables. Estimated values for β_0 to β_5 are shown in Table 8.6, together with estimated standard errors and chi-squared statistics for testing whether the individual estimates differ significantly from zero. Again, there is no evidence of any significant effects.

The effect of adding X_1^2 to X_5^2 to the model was also investigated. This did not introduce any significant results. Adding the ten product terms $X_1 X_2$, $X_1 X_3$, ..., $X_4 X_5$ as well as the squared terms was also investigated, but in this case the computer program being used failed to fit the logistic function, probably because there were then 21 parameters to be estimated using only 49 data points.

In summary, logistic regression gives no indication at all that the survival of the female sparrows was related to the measured variables.

Table 8.6 Estimates of the Constant Term and the Coefficients of X Variables when a Logistic Regression Model Is Fitted to Data on the Survival of 49 Female Sparrows

Variable	β estimate	Standard error	Chi-squared	P-value
Constant	13.582	15.865	—	—
Total length	–0.163	0.140	1.36	0.244
Alar extent	–0.028	0.106	0.07	0.794
Length beak and head	–0.084	0.629	0.02	0.894
Length humerus	1.062	1.023	1.08	0.299
Length keel of sternum	0.072	0.417	0.03	0.864

Note: The chi-squared value is (estimate/standard error)². The p-value is the probability of a value this large from the chi-squared distribution with one degree of freedom. A small p-value (say less than 0.05) provides evidence that the true value of the param- eter concerned is not equal to zero.

Example 8.4   Comparison of two samples of Egyptian skulls

As an example of separate sampling, where the sample size in the two groups being compared is not necessarily related in any way to the respective population sizes, consider the comparison between the first and last samples of Egyptian skulls for which data are provided in Table 1.2. The first sample consists of 30 male skulls from burials in the area of Thebes during the early predynastic period (circa 4000 B.C.) in Egypt, and the last sample consists of 30 male skulls from burials in the same area during the Roman period (circa A.D. 150). For each skull, measurements are available for X_1 = maximum breadth, X_2 = basibregmatic height, X_3 = basialveolar length, and X_4 = nasal height, all in mm (Figure 1.1). For the purpose of this example, it will be assumed that the two samples were effectively randomly chosen from their respective populations, although there is no way of knowing how realistic this is.

Obviously, the equal sample sizes in no way indicate that the population sizes in the two periods were equal. The sizes are in fact completely arbitrary because many more skulls have been measured from both periods, and an unknown number of skulls have either not survived intact or have not been found. Therefore, if the two samples are lumped together and treated as a sample of size 60 for the estimation of a logistic regression equation, then it is clear that the estimated probability of a skull with certain dimensions being from the early predynastic period may not really be estimating the true probability at all.

In fact, it is difficult to define precisely what is meant by the true probability in this example because the population is not at all clear. A working definition is that the probability of a skull with specified dimensions being from the predynastic period is equal to the proportion of all skulls with the given dimensions that are from the predynastic period in a hypothetical population of all male skulls from either the predynastic or the Roman period that might have been recovered by archaeologists in the Thebes region.

It can be shown (Seber, 1984, p. 312) that if a logistic regression is carried out on a lumped sample to estimate the value obtained in Equation 8.3, then the modified equation

$$\pi_i = \frac{\exp\left(\beta_0 - \log_e\left\{(n_1 P_2)/(n_2 P_1)\right\} + \beta_1 x_{i1} + \beta_2 x_{i2} + \ldots + \beta_p x_{ip}\right)}{1 + \exp\left(\beta_0 - \log_e\left\{(n_1 P_2)/(n_2 P_1)\right\} + \beta_1 x_{i1} + \beta_2 x_{i2} + \ldots + \beta_p x_{ip}\right)} \quad (8.4)$$

is what really gives the probability that an item with the specified X values is a success. Here Equation 8.4 differs from Equation 8.3 because of the term $\log_e\{(n_1 P_2)/(n_2 P_1)\}$ in the numerator and the denominator, where P_1 is the proportion of items in the full population of successes and failures that are successes, and $P_2 = 1 - P_1$ is the proportion of the population that are failures. This then means that in order to estimate the probability of an item with the specified X values being a success, the values for P_1 and P_2 must either be

Table 8.7 Estimates of the Constant Term and the Coefficients of X Variables When a Logistic Regression Model Is Fitted to Data on 30 Predynastic and 30 Roman Period Male Egyptian skulls

Variable	β estimate	Standard error	Chi-squared	P-value
Constant	−6.732	13.081	—	—
Maximum breadth	−0.202	0.075	7.13	0.008
Basibregmatic height	0.129	0.079	2.66	0.103
Basialveolar length	0.177	0.073	5.84	0.016
Nasal height	−0.008	0.104	0.01	0.939

Note: The chi-squared value is (estimated β error/standard error)². The p-value is the probability of a value this large from the chi-squared distribution with one degree of freedom. A small p-value (say less than 0.05) provides evidence that the true value of the parameter concerned is not equal to zero.

known or can somehow be estimated separately from the sample data, in order to adjust the estimated logistic regression equation for the fact that the sample sizes n_1 and n_2 are not proportional to the population frequencies of successes and failures. In the example being considered, this requires that estimates of the relative frequencies of predynastic and Roman skulls in the Thebes area must be known in order to be able to estimate the probability that a skull is predynastic based on the values that it possesses for the variables X_1 to X_4.

Logistic regression was applied to the lumped data from the 60 predynastic and Roman skulls, with a predynastic skull being treated as a success. The resulting chi-squared test for the extent to which success is related to the X variables is 27.13 with four degrees of freedom. This is significantly large at the 0.1% level, giving very strong evidence of a relationship. The estimates of the constant term and the coefficients of the X variables are shown in Table 8.7. It can be seen that the estimate of β_1 is significantly different from zero at about the 1% level and that β_3 is significantly different from zero at the 2% level. Hence X_1 and X_3 appear to be the important variables for discriminating between the two types of skull.

The fitted function can be used to discriminate between the two groups of skulls by assigning values for P_1 and $P_2 = 1 - P_1$ in Equation 8.4. As already noted, it is desirable that these should correspond to the population proportions of predynastic and Roman skulls. However, this is not possible because these proportions are not known. In practice, therefore, arbitrary values must be assigned. For example, suppose that P_1 and P_2 are both set equal to 0.5. Then $\log_e\{(n_1 P_2)/(n_2 P_1)\} = \log_e(1) = 0$, because $n_1 = n_2$, and Equation 8.3 and Equation 8.4 become identical. The logistic function therefore estimates the probability of a skull's being predynastic in a population with equal frequencies of predynastic and Roman skulls.

The extent to which the logistic equation is effective for discrimination is indicated in Figure 8.2, which shows the estimated values of π_i for the 60 sample skulls. There is a distinct difference in the distributions of values for the two samples, with the mean for predynastic skulls being about 0.7 and

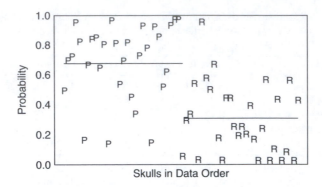

Figure 8.2 Values from the fitted logistic regression function plotted for 30 predy-nastic (P) and 30 Roman (R) skulls. The horizontal lines indicate the average group probabilities.

the mean for Roman skulls being about 0.3. However, there is also a consid-erable overlap between the distributions. As a result, if the sample skulls are classified as being predynastic when the logistic equation gives a value greater than 0.5, or as Roman when the equation gives a value of less than 0.5, then six predynastic skulls are misclassified as being Roman, and seven Roman skulls are misclassified as being predynastic.

8.11   Computer programs

The calculations for the examples used in this chapter were carried out using the program NCSS (Hintze, 2001). However, major statistical packages gen-erally have a discriminant function option that applies the methods described in Sections 8.2 to 8.5, based on the assumption of normally dis-tributed data. Because the details of the order of calculations, the way the output is given, and the terminology vary considerably, manuals may have to be studied carefully to determined precisely what is done by any individ-ual program. Logistic regression is also fairly widely available. In some programs, there is the restriction that all items are assumed to have different values for X variables. However, it is more common for groups of items with common X values to be permitted.

8.12   Discussion and further reading

The assumption that samples are from multivariate distributions with the same covariance matrix that is required for the use of the methods described in Sections 8.2 to 8.5 can sometimes be relaxed. If the samples being com-pared are assumed to come from multivariate normal distributions with different covariance matrices, then a method called quadratic discriminant function analysis can be applied. This option is also available in many com-puter packages. See Seber (1984, p. 297) for more information about this method and a discussion of its performance relative to the more standard linear discriminant function analysis.

Discrimination using logistic regression has been described in Section 8.10 in terms of the comparison of two groups. More detailed treatments of this method are provided by Hosmer and Lemeshow (2000) and Collett (1991). The method can also be generalized for discrimination between more than two groups, if necessary, under several names, including polytomous regression. See Hosmer and Lemeshow (2000, ch. 8) for more details. This type of analysis is now becoming a standard option in computer packages.

8.13   Chapter summary

- The problem of separating individuals in different groups based on the measurements that the individuals have for p variables is described.
- One solution to this problem involves allocating each individual to the group that it is closest to in terms of the Mahalanobis distance, and then seeing what proportion of individuals are misclassified.
- An alternative approach ascribed to Fisher (1936) is based on the idea of finding the linear combination of the p variables that maximizes the differences between the groups in the sense of the F-statistic from an analysis of variance. This approach leads to s linear combinations, where s is the minimum of the number of variables and the number of groups minus one. Finding these linear combinations is an eigenvalue problem. The linear combinations are uncorrelated within groups.
- Tests of significance to determine how many linear combinations are needed to describe group differences are discussed. Some tests that are commonly used may not provide good results.
- The assumptions of standard discriminant function analysis (normality and equal within-group covariance matrices) are discussed.
- Two examples are considered involving the comparison of five samples of male Egyptian skulls from the predynastic and Roman periods, and employment patterns in four groups of European countries.
- The allowance for groups with different sizes is discussed, where a random individual is not equally likely to be in any group.
- Stepwise discriminant function analysis (with the stepwise selection of variables to use in the function) is discussed, with a randomization-based method for testing the properties of the method.
- The jackknife method for estimating the probabilities of correctly allocating individuals to groups is described.
- The problem of assigning ungrouped individuals to groups is discussed.

- Logistic regression is described as an alternative to the usual method of discriminant function analysis based on the assumption of normally distributed data. Three types of sampling schemes are also described.
- The logistic regression approach to discrimination is illustrated using the examples of discriminating between female sparrows that did or did not survive a severe storm, and the comparison of two samples of male Egyptian skulls.
- Computer programs for standard discriminant function analysis and logistic regression are discussed.
- Extensions to the methods covered in the chapter are described. These extensions allow the methods to be applied to groups that have different covariance matrices, and they allow the logistic regression type of approach to be applied to more than two groups.

Exercises

1. Consider the data in Table 4.5 for nine mandible measurements on samples from five different canine groups. Carry out a discriminant function analysis to see how well it is possible to separate the groups using the measurements.
2. Still considering the data in Table 4.5, investigate each canine group separately to see whether logistic regression shows a significant difference between males and females for the measurements. Note that in view of the small sample sizes available for each group, it is unreasonable to expect to fit a logistic function involving all nine variables with good estimates of parameters. Therefore, consideration should be given to fitting functions using only a subset of the variables.

References

Collett, D. (1991), *Modelling Binary Data*, Chapman and Hall, London.

Fisher, R.A. (1936), The utilization of multiple measurements in taxonomic problems, *Ann. Eugenics*, 7, 179–188.

Harris, R.J. (1985), *A Primer on Multivariate Statistics*, 2nd ed., Academic Press, Orlando, FL.

Hintze, J. (2001), NCSS and PASS, Number Cruncher Statistical Systems, Kaysville, Utah, www.ncss.com.

Hosmer, D.W. and Lemeshow, S. (2000), *Applied Logistic Regression*, 2nd ed., Wiley, New York.

Manly, B.F.J. (1997), *Randomization, Bootstrap and Monte Carlo Methods in Biology*, 2nd ed., Chapman and Hall, London.

Seber, G.A.F. (1984), *Multivariate Observations*, Wiley, New York.

chapter nine

Cluster analysis

9.1 Uses of cluster analysis

Suppose that there is a sample of n objects, each of which has a score on p variables. Then the idea with a cluster analysis is to use the values of the variables to devise a scheme for grouping the objects into classes so that similar objects are in the same class. The method used must be completely numerical, and the number of classes is not usually known. This problem is clearly more difficult than the problem for a discriminant function analysis that was considered in the last chapter, because with discriminant function analysis, the groups are known to begin with.

There are many reasons why cluster analysis may be worthwhile. It might be a question of finding the true groups that are assumed to really exist. For example, in psychiatry there has been disagreement over the classification of depressed patients, and cluster analysis has been used to define objective groups. Cluster analysis may also be useful for data reduction. For example, a large number of cities can potentially be used as test markets for a new product, but it is only feasible to use a few. If cities can be placed into a small number of groups of similar cities, then one member from each group can be used for the test market. Alternatively, if cluster analysis generates unexpected groupings, then this might in itself suggest relationships to be investigated.

9.2 Types of cluster analysis

Many algorithms have been proposed for cluster analysis. Here, attention will mostly be restricted to those following two particular approaches. First, there are hierarchic techniques that produce a dendrogram, as shown in Figure 9.1. These methods start with the calculation of the distances of each object to all other objects. Groups are then formed by a process of agglomeration or division. With agglomeration, all objects start by being alone in groups of one. Close groups are then gradually merged until finally all objects are in a single group. With division, all objects start in a single group.

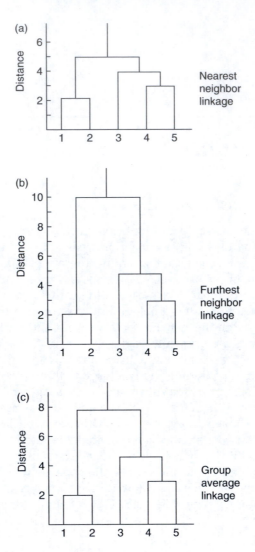

Figure 9.1 Examples of dendrograms from cluster analyses of five objects.

This is then split into two groups, the two groups are then split, and so on until all objects are in groups of their own.

The second approach to cluster analysis involves partitioning, with objects being allowed to move in and out of groups at different stages of the analysis. There are many variations on the algorithm used, but the basic approach involves first choosing some more or less arbitrary group centers, with objects then allocated to the nearest one. New centers are then calculated where these represent the averages of the objects in the groups. An object is then moved to a new group if it is closer to that group's center than it is to the center of its present group. Any groups that are close together are merged, spread-out groups are split, etc., following some defined rules. The

Table 9.1 A Matrix Showing the Distances between
Five Objects

Object	Object				
	1	2	3	4	5
1	—				
2	2	—			
3	6	5	—		
4	10	9	4	—	
5	9	8	5	3	—

Note: The distance is always zero between an object and
itself, and the distance from object i to object j is the
same as the distance from object j to object i.

Table 9.2 The Merging of Groups
Based on Nearest-Neighbor Distances

Distance	Groups
0	1, 2, 3, 4, 5
2	(1,2), 3, 4, 5
3	(1,2), 3, (4,5)
4	(1,2), (3,4,5)
5	(1,2,3,4,5)

process continues iteratively until stability is achieved with a predetermined
number of groups. Usually a range of values is tried for the final number of
groups.

9.3 Hierarchic methods

Agglomerative hierarchic methods start with a matrix of distances between
objects. All objects begin alone in groups of size one, and groups that are
close together are merged. There are various ways to define close. The
simplest is in terms of nearest neighbors. For example, suppose that the
distances between five objects are as shown in Table 9.1. The calculations for
potential groupings are then as shown in Table 9.2.

Groups are merged at a given level of distance if one of the objects in
one group is that distance or closer to at least one object in the second group.
At a distance of 0, all five objects are on their own. The smallest distance
between two objects is 2, which is between the first and second objects. Hence
at a distance level of 2, there are four groups (1, 2), (3), (4), and (5). The next
smallest distance between objects is 3, which is between objects 4 and 5.
Hence at a distance of 3, there are three groups (1,2), (3), and (4,5). The next
smallest distance is 4, which is between objects 3 and 4. Hence at this level
of distance, there are two groups (1,2) and (3,4,5). Finally, the next smallest

Table 9.3 The Merging of Groups Based
on Furthest-Neighbor Distances

Distance	Groups
0	1, 2, 3, 4, 5
2	(1,2), 3, 4, 5
3	(1,2), 3, (4,5)
5	(1,2), (3,4,5)
10	(1,2,3,4,5)

Table 9.4 The Merging of Groups Based
on Group Average Distances

Distance	Groups
0	1, 2, 3, 4, 5
2	(1,2), 3, 4, 5
3	(1,2), 3, (4,5)
4.5	(1,2), (3,4,5)
7.8	(1,2,3,4,5)

distance is 5, which is between objects 2 and 3 and between objects 3 and 5. At this level, the two groups merge into the single group (1,2,3,4,5), and the analysis is complete. The dendrogram shown in Figure 9.1(a) illustrates how the agglomeration takes place.

With furthest-neighbor linkage, two groups merge only if the most distant members of the two groups are close enough. With the example data, this works as shown in Table 9.3. Object 3 does not join with objects 4 and 5 until distance level 5 because this is the distance to object 3 from the furthest away objects 4 and 5. The furthest-neighbor dendrogram is shown in Figure 9.1(b).

With group average linkage, two groups merge if the average distance between them is small enough. With the example data, this gives the results shown in Table 9.4. For example, groups (1,2) and (3,4,5) merge at distance level 7.8, as this is the average distance from objects 1 and 2 to objects 3, 4, and 5, the actual distances being 1 to 3 = 6; 1 to 4 = 10; 1 to 5 = 9; 2 to 3 = 5; 2 to 4 = 9; and 2 to 5 = 8, with $(6 + 10 + 9 + 5 + 9 + 8)/6 = 7.8$. The dendrogram in this case is shown in Figure 9.1(c).

Divisive hierarchic methods have been used less often than agglomerative ones. The objects are all put into one group initially, and then this is split into two groups by separating off the object that is furthest on average from the other objects. Objects from the main group are then moved to the new group if they are closer to this group than they are to the main group. Further subdivisions occur as the distance that is allowed between objects in the same group is reduced. Eventually all objects are in groups of size one.

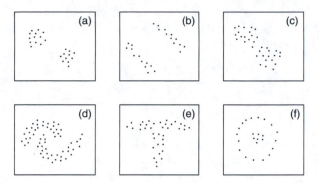

Figure 9.2 Some possible patterns of points when there are two clusters.

9.4   Problems of cluster analysis

It has already been mentioned that there are many algorithms for cluster analysis. However, there is no generally accepted best method. Unfortunately, different algorithms do not necessarily produce the same results on a given set of data, and there is usually rather a large subjective component in the assessment of the results from any particular method.

A fair test of any algorithm is to take a set of data with a known group structure and see whether the algorithm is able to reproduce this structure. It seems to be the case that this test works only in cases where the groups are very distinct. When there is a considerable overlap between the initial groups, a cluster analysis may produce a solution that is quite different from the true situation.

In some cases, difficulties will arise because of the shape of clusters. For example, suppose that there are two variables X_1 and X_2, and objects are plotted according to their values for these. Some possible patterns of points are illustrated in Figure 9.2. Case (a) is likely to be found by any reasonable algorithm, as is case (b). In case (c), some algorithms might well fail to detect two clusters because of the intermediate points. Most algorithms would have trouble handling cases like (d), (e), and (f).

Of course, clusters can be based only on the variables that are given in the data. Therefore, they must be relevant to the classification wanted. To classify depressed patients, there is presumably not much point in measuring height, weight, or length of arms. A problem here is that the clusters obtained may be rather sensitive to the particular choice of variables that is made. A different choice of variables, apparently equally reasonable, may give different clusters.

9.5   Measures of distance

The data for a cluster analysis usually consist of the values of p variables $X_1, X_2, \ldots, X_p$ for n objects. For hierarchic algorithms, these values are then used to produce an array of distances between the objects. Measures of distance have already been discussed in Chapter 5. Here it suffices to say that the Euclidean distance function

$$d_{ij} = \left\{ \sum_{k=1}^{P} \left(x_{ik} - x_{jk} \right)^2 \right\}^{1/2} \qquad (9.1)$$

is often used, where x_{ik} is the value of variable X_k for object i and x_{jk} is the value of the same variable for object j. The geometrical interpretation of the distance d_{ij} is illustrated in Figures 5.1 and 5.2 for the cases of two and three variables.

Usually variables are standardized in some way before distances are calculated, so that all p variables are equally important in determining these distances. This can be done by coding the variables so that the means are all zero and the variances are all one. Alternatively, each variable can be coded to have a minimum of zero and a maximum of one. Unfortunately, standardization has the effect of minimizing group differences, because if groups are separated well by the variable X_i, then the variance of this variable will be large. In fact, it should be large. It would be best to be able to make the variances equal to one within clusters, but this is obviously not possible, as the whole point of the analysis is to find the clusters.

9.6   Principal components analysis with cluster analysis

Some cluster analysis algorithms begin by doing a principal components analysis to reduce a large number of original variables down to a smaller number of principal components. This can drastically reduce the computing time for the cluster analysis. However, it is known that the results of a cluster analysis can be rather different with and without the initial principal components analysis. Consequently, an initial principal components analysis is probably best avoided, because computing time is seldom an issue in the present day.

On the other hand, when the first two principal components account for a high percentage of variation in the data, a plot of individuals against these two components is certainly a useful way of looking for clusters. For example, Figure 6.2 shows European countries plotted in this way for principal components based on employment percentages. The countries do seem to group in a meaningful way.

Example 9.1   Clustering of European countries

The data just mentioned on the percentages of people employed in nine industry groups in different countries of Europe (Table 1.5) can be used for a first example of cluster analysis. The analysis should show which countries have similar employment patterns and which countries are different in this respect. As shown in Table 1.5, a sensible grouping existed when the data were collected, consisting of (1) the European Union (EU) countries, (2) the European Free Trade Area (EFTA) countries, (3) the eastern European countries, and (4) the four other countries of Cyprus, Gibraltar, Malta, and Turkey. It is therefore interesting to see whether this grouping can be recovered using a cluster analysis.

The first step in the analysis involves standardizing the nine variables so that each one has a mean of zero and a standard deviation of one. For example, variable 1 is AGR, the percentage employed in agriculture, forestry, and fishing. For the 30 countries being considered, this variable has a mean of 12.19 and a standard deviation of 12.31, with the latter value calculated using Equation 4.1. The AGR data value for Belgium is 2.6, which standardizes to $(2.6 - 12.19)/12.31 = -0.78$. Similarly, the data value for Denmark is 5.6, which standardizes to −0.54, and so on. The standardized data values are shown in Table 9.5.

The next step in the analysis involves calculating the Euclidean distances between all pairs of countries. This can be done by applying Equation 9.1 to the standardized data values. Finally, a dendrogram can be formed using, for example, the agglomerative, nearest-neighbor, hierarchic process described above. In practice, all of these steps can be carried out using a suitable statistical package.

The dendrogram obtained using the NCSS package (Hintze, 2001) is shown in Figure 9.3. This package actually uses a slightly different definition of the Euclidean distance than the one defined by Equation 9.1. In effect, it uses d_{ij} as defined by that equation divided by the square root of the number of variables (p). However, this makes no difference to the shape of the dendrogram and merely changes the horizontal axis by a constant factor.

It can be seen that the two closest countries were Sweden and Denmark. These are at a distance of about 0.4 apart. At a slightly larger distance, Belgium joins these two countries to make a cluster of size three. As the distance increases, more and more countries combine, and the amalgamation ends with Albania joining all of the other countries in one cluster, at a distance of about 1.7.

One interpretation of the dendrogram is that there are just four clusters defined by a nearest-neighbor distance of about 1.0. These are then (1) Albania, (2) Hungary and the Czech/Slovak Republics, (3) Gibraltar, and (4) all of the other countries. This then separates off three eastern countries and Gibraltar from everything else, which suggests that the classification into EU, EFTA, eastern, and other countries is not a good indicator of employment patterns. This contradicts the reasonably successful separation of eastern and EU/EFTA countries from a discriminant function analysis (Figure 8.1). However, there is some limited agreement with the plot of countries against the first two principal components, where Albania and Gibraltar show up as having very extreme data values (Figure 6.2).

An alternative analysis was carried out using the K-means clustering option in the NCSS package (Hintze, 2001). This essentially uses the

Table 9.5 Standardized Values for Percentages Employed in Different Industry Groups in Europe

Country	AGR	MIN	MAN	PS	CON	SER	FIN	SPS	TC
Belgium	-0.78	-0.37	0.05	0.00	-0.45	0.24	0.51	1.13	0.28
Denmark	-0.54	-0.38	0.01	-0.16	-0.41	-0.22	0.61	1.07	0.44
France	-0.58	-0.35	-0.01	0.16	-0.16	0.21	0.89	0.70	-0.04
Germany	-0.73	-0.31	0.48	0.32	0.68	0.30	0.74	0.16	-0.69
Greece	0.81	-0.33	-0.11	0.32	-0.27	0.50	-0.34	-0.82	0.36
Ireland	0.13	-0.32	-0.05	0.64	-0.16	0.42	0.44	-0.17	-0.53
Italy	-0.31	-0.26	0.17	-1.29	0.57	1.16	-0.51	0.12	-0.94
Luxembourg	-0.72	-0.38	-0.07	-0.16	0.87	1.08	0.51	0.30	0.28
Netherlands	-0.65	-0.38	-0.11	-0.16	-2.54	0.55	1.22	1.29	0.28
Portugal	-0.06	-0.33	0.35	-0.16	0.25	0.81	-0.09	-0.27	-1.34
Spain	-0.19	-0.33	0.09	-0.32	0.72	0.86	-0.19	-0.03	-0.53
U.K.	-0.81	-0.31	0.11	0.64	-0.19	0.88	1.44	0.16	0.04
Austria	-0.39	-0.35	0.70	0.64	0.35	0.67	0.01	-0.42	-0.04
Finland	-0.30	-0.37	-0.10	0.64	-0.27	-0.20	0.49	0.71	0.85
Iceland	-0.14	-0.39	-0.17	0.16	0.90	-0.22	0.34	0.42	0.20
Norway	-0.52	-0.26	-0.60	0.48	-0.38	0.38	0.24	1.20	1.34
Sweden	-0.73	-0.35	-0.14	0.00	-0.41	-0.28	0.69	1.43	0.61
Switzerland	-0.54	-0.39	0.47	-1.29	0.61	0.94	1.02	-0.45	-0.21
Albania	3.52	1.80	-2.15	-1.29	-1.51	-2.39	2.17	-3.09	-2.80
Bulgaria	0.55	-0.39	1.56	-1.29	-0.30	-1.21	-1.29	-0.70	0.85
Czech/Slovak Reps	0.05	3.82	-2.15	-1.29	0.32	-1.05	-1.27	-0.47	0.36
Hungary	0.25	2.87	-2.15	-1.29	-0.41	-0.45	-1.67	0.04	1.90
Poland	0.93	0.05	0.40	0.16	-0.45	-1.03	-1.34	-0.29	-1.02
Romania	0.80	-0.10	1.86	1.93	-0.63	-1.69	-1.52	-1.34	0.28
USSR (Former)	0.51	-0.39	0.90	-1.29	0.98	-1.50	-1.52	-0.16	1.58
Yugoslavia (Former)	-0.58	-0.14	1.95	2.25	0.21	-0.36	-0.89	-0.90	1.09
Cyprus	0.11	-0.35	-0.14	-0.48	0.57	1.56	0.01	-0.66	-0.37
Gibraltar	-0.99	-0.39	-1.43	1.93	3.43	1.72	1.04	0.80	-1.18
Malta	-0.78	-0.32	0.81	1.13	-1.07	-1.05	-0.69	1.67	0.61
Turkey	2.65	-0.29	-0.53	-0.97	-0.85	-0.63	-1.07	-1.43	-1.66

Note: Derived from the percentages in Table 1.5.

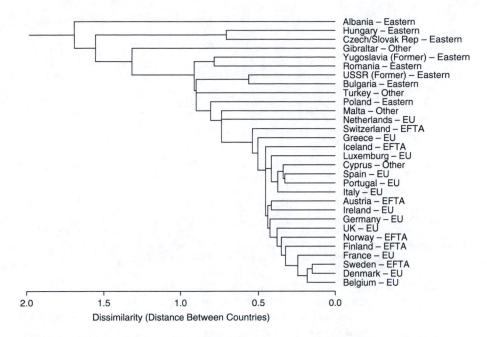

Albania – Eastern
Hungary – Eastern
Czech/Slovak Rep – Eastern
Gibraltar – Other
Yugoslavia (Former) – Eastern
Romania – Eastern
USSR (Former) – Eastern
Bulgaria – Eastern
Turkey – Other
Poland – Eastern
Malta – Other
Netherlands – EU
Switzerland – EFTA
Greece – EU
Iceland – EFTA
Luxemburg – EU
Cyprus – Other
Spain – EU
Portugal – EU
Italy – EU
Austria – EFTA
Ireland – EU
Germany – EU
UK – EU
Norway – EFTA
Finland – EFTA
France – EU
Sweden – EFTA
Denmark – EU
Belgium – EU

2.0   1.5   1.0   0.5   0.0
Dissimilarity (Distance Between Countries)

Figure 9.3 The dendrogram obtained from a nearest-neighbor, hierarchical cluster analysis on employment data from European countries.

partitioning method described in Section 9.2, which starts with arbitrary cluster centers, allocates items to the nearest center, recalculates the mean values of variables for each group, reallocates individuals to their closest group centers to minimize the within-cluster total sum of squares, and so on. The calculations use standardized variables with means of zero and standard deviations of one. Ten random choices of starting clusters were tried, with from two to six clusters.

The percentage of the variation accounted for ranged from 73.5% with two clusters to 27.6% with six clusters. With four clusters, these were (1) Turkey and Albania, (2) Hungary and the Czech/Slovak Republics, (3) Bulgaria, Poland, Romania, the USSR (former), Yugoslavia (former),and Malta, and (4) the EU and EFTA countries, Cyprus, and Gibraltar. This is not the same as the four-cluster solution given by the dendrogram of Figure 9.3, although there are some similarities. No doubt other algorithms for cluster analysis will give slightly different solutions.

Example 9.2   Relationships between canine species

As a second example, consider the data provided in Table 1.4 for mean mandible measurements of seven canine groups. As has been explained before, these data were originally collected as part of a study on the relationship between prehistoric dogs, whose remains have been uncovered in Thailand, and the other six living species. This question has already been

Table 9.6 Clusters Found at Different Distance Levels for a Hierarchic Nearest-Neighbor Cluster Analysis

Distance	Cluster	Number of Clusters
0.00	MD, PD, GJ, CW, IW, CU, DI	6
0.72	(MD, PD), GJ, CW, IW, CU, DI	5
1.38	(MD, PD, CU), GJ, CW, IW, DI	5
1.63	(MD, PD, CU), GJ, CW, IW, DI	5
1.68	(MD, PD, CU, DI), GJ, CW, IW	4
1.80	(MD, PD, CU, DI), GJ, CW, IW	4
1.84	(MD, PD, CU, DI), GJ, CW, IW	4
2.07	(MD, PD, CU, DI, GJ), CW, IW	3
2.31	(MD, PD, CU, DI, GJ), (CW, IW)	2
2.37	(MD, PD, CU, DI, GJ, CW, IW)	1

Note: MD = modern dog, GJ = golden jackal, CW = Chinese wolf, IW = Indian wolf, CU = cuon, DI = dingo, and PD = prehistoric dog.

considered in terms of distances between the seven groups in Example 5.1. Table 5.1 shows mandible measurements standardized to have means of zero and standard deviations of one. Table 5.2 shows Euclidean distances between the groups based on these standardized measurements.

With only seven species to cluster, it is a simple matter to carry out a nearest-neighbor, hierarchic cluster analysis without using a computer. Thus it can be seen from Table 5.2 that the two most similar species are the prehistoric dog and the modern dog, at a distance of 0.72. These therefore join into a single cluster at that level. The next largest distance is 1.38 between the cuon and the prehistoric dog, so that at that level, the cuon joins the cluster with the prehistoric and modern dog. The third largest distance is 1.63 between the cuon and modern dog, but because these are already in the same cluster, this has no effect. Continuing in this way produces the clusters that are shown in Table 9.6. The corresponding dendrogram is shown in Figure 9.4.

It appears that the prehistoric dog is closely related to the modern Thai dog, with both of these being somewhat related to the cuon and dingo and less closely related to the golden jackal. The Indian and Chinese wolves are closest to each other, but the difference between them is relatively large.

It seems fair to say that in this example, the cluster analysis has produced a sensible description of the relationship between the different groups.

9.7 *Computer programs*

Computer programs for cluster analysis are widely available, and the larger statistical packages often include a variety of different options for both hierarchic and partitioning methods. As the results obtained usually vary to some extent, depending on the precise details of the algorithms used, it will

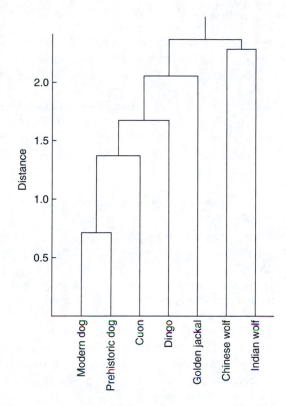

Figure 9.4 The dendrogram obtained from a nearest-neighbor cluster analysis for the relationship between canine species.

usually be worthwhile to try several options before deciding on the final method to be used for an analysis.

9.8   Discussion and further reading

A number of books devoted to cluster analysis are available, including the classic texts by Hartigan (1975) and Romesburg (1984), and the more recent one by Everitt et al. (2001).

An approach to clustering that has not been considered in this chapter involves assuming that the data available come from a mixture of several different populations for which the distributions are of a type that is assumed to be known (e.g., multivariate normal). The clustering problem is then transformed into the problem of estimating, for each of the populations, the parameters of the assumed distribution and the probability that an observation comes from that population. This approach has the merit of moving the clustering problem away from the development of *ad hoc* procedures towards the more usual statistical framework of parameter estimation and model testing. See Everitt et al. (2001, ch. 6) for an introduction to this method.

9.9   Chapter summary

- The reasons for carrying out a cluster analysis are discussed. These include defining true underlying groups and finding a small number of objects (one per cluster) that cover the full set of conditions for a larger set of objects.
- Two types of cluster analysis are described. One results in a dendrogram showing a hierarchic relationship between objects depending on their similarities. The other involves an iterative partitioning procedure to find the best set of n clusters for a set of data by starting with arbitrary clusters and improving these by moving individuals between them.
- There are a variety of agglomerative, hierarchic clustering algorithms. Those based on nearest-neighbor distances, furthest-neighbor distances, and group averages are described. These start with all individuals in groups on their own and gradually merge them into one group.
- Divisive hierarchic methods are also briefly described, although these are not used as often as agglomerative methods. The divisive methods start with all objects in one group and gradually separate the objects until each one is in a group on its own.
- Problems in detecting clusters with unusual shapes are discussed.
- Measures of the distance between objects are discussed, and the Euclidean distance in particular. The need for standardization of variables is also mentioned.
- Cluster analysis using principal components is sometimes used to reduce computing times. It is suggested that this is probably best avoided.
- An example is provided of a cluster analysis on European countries, based on the percentages in different employment groups. Results are shown based on agglomerative, hierarchic clustering, and also a partitioning method called K-means clustering.
- A second example considers the relationship between prehistoric dogs from Thailand and six existing canine species.
- Computer programs for carrying out cluster analysis are briefly discussed.
- Sources of further information about cluster analysis are described, and some alternative approaches to clustering are identified.

Exercises

Exercise 1

Table 9.7 shows the abundances of the 25 most abundant plant species on 17 plots from a grazed meadow in Steneryd Nature Reserve in Sweden as measured by Persson (1981) and used for an example by Digby and Kempton (1987). Each value in the Table is the sum of cover values in a range from 0 to 5 for nine sample quadrants, so that a value of 45 corresponds to the complete cover by the species being considered. Note that the species are in order from the most abundant (1) to the least abundant (25), and the plots are in the order given by Digby and Kempton (1987, Table 3.2), which corresponds to variation in certain environmental factors such as light and moisture. Carry out cluster analyses to study the relationships between (a) the 17 plots and (b) the 25 species.

Exercise 2

Table 9.8 shows a set of data concerning grave goods from a cemetery at Bannadi, northeast Thailand, that were kindly supplied by Professor C.F.W. Higham. These data consist of a record of the presence or absence of 38 different types of article in each of 47 graves, with additional information on whether the remains were of an adult male, adult female, or a child. The burials are in the order of richness of the different types of goods (totals ranging from 0 to 11), and the goods are in the order of the frequency of occurrence (totals ranging from 1 to 18). Carry out a cluster analysis to study the relationships between the 47 burials. Is there any clustering in terms of the type of remains?

Table 9.7 Abundance Measures for 25 Plant Species on 17 Plots in Steneryd Nature Reserve, Sweden

Species	1	2	3	4	5	6	7	8	9	10	11	12	13	14	15	16	17
Festuca ovina	38	43	43	30	10	11	20	0	0	5	4	1	1	0	0	0	0
Anemone nemorosa	0	0	0	4	10	7	21	14	13	19	20	19	6	10	12	14	21
Stallaria holostea	0	0	0	0	0	6	8	21	39	31	7	12	0	16	11	6	9
Agrostis tenuis	10	12	19	15	16	9	0	9	28	8	0	4	0	0	0	0	0
Ranunculus ficaria	0	0	0	0	0	0	0	0	0	0	13	0	0	21	20	21	37
Mercurialis perennis	0	0	0	0	0	0	0	0	0	0	1	0	0	0	11	45	45
Poa pratenis	1	0	5	6	2	8	10	15	12	15	4	5	6	7	0	0	0
Rumex acetosa	0	7	0	10	9	9	3	9	8	9	2	5	5	1	7	0	0
Veronica chamaedrys	0	0	1	4	6	9	9	9	11	11	6	5	4	1	7	0	0
Dactylis glomerata	0	0	0	0	0	8	0	14	2	14	3	9	8	7	7	2	1
Fraxinus excelsior (juv.)	0	0	0	0	0	8	0	1	6	5	4	7	9	8	8	7	6
Saxifraga granulata	0	5	3	9	12	9	0	1	7	4	5	1	1	1	3	0	0
Deschampsia flexuosa	0	0	0	0	0	0	30	0	14	3	8	0	3	3	0	0	0
Luzula campestris	4	10	10	9	7	6	9	0	0	2	1	0	2	0	1	0	0
Plantago lanceolata	2	9	7	15	13	8	0	0	0	0	0	0	0	0	0	0	0
Festuca rubra	0	0	0	0	15	6	0	18	1	9	0	0	2	0	0	0	0
Hieracium pilosella	12	7	16	8	1	6	0	0	0	0	0	0	0	0	0	0	0
Geum urbanum	0	0	0	0	0	7	0	2	2	1	0	7	9	2	3	8	7
Lathyrus montanus	0	0	0	0	0	7	9	2	12	6	3	8	0	0	0	0	0
Campanula persicifolia	0	0	0	2	2	6	3	0	6	5	3	9	3	2	7	0	0
Viola riviniana	0	0	0	0	0	4	1	4	2	9	6	8	4	1	6	0	0
Hepatica nobilis	0	0	0	0	0	8	0	0	0	6	2	10	6	0	2	7	0
Achillea millefolium	1	9	16	9	5	2	0	4	0	0	0	0	0	0	0	0	0
Allium sp.	0	0	0	0	2	7	0	0	0	3	1	6	8	2	0	7	4
Trifolim repens	0	0	6	14	19	2	0	0	0	0	0	0	0	0	0	0	0

Table 9.8 Grave Goods in Burials in the Bannadi Cemetery in Northern Thailand

Burial	Type	1	2	3	4	5	6	7	8	9	10	11	12	13	14	15	16	17	18	19	20	21	22	23	24	25	26	27	28	29	30	31	32	33	34	35	36	37	38	Sum
B33	3	0	0	0	0	0	0	0	0	0	0	0	0	0	0	0	0	0	0	0	0	0	0	0	0	0	0	0	0	0	0	0	0	0	0	0	0	0	0	0
B9	2	0	0	0	0	0	0	0	0	0	0	0	0	0	0	0	0	0	0	0	0	0	0	0	0	0	0	0	0	0	0	0	0	0	0	0	0	0	0	0
B32	2	0	0	0	0	0	0	0	0	0	0	0	0	0	0	0	0	0	0	0	0	0	0	0	0	0	0	0	0	0	0	0	0	0	0	0	0	0	0	0
B11	1	0	0	0	0	0	0	0	0	0	0	0	0	0	0	0	0	0	0	0	0	0	0	0	0	0	0	0	0	0	0	0	0	0	0	0	0	0	0	0
B28	1	0	0	0	0	0	0	0	0	0	0	0	0	0	0	0	0	0	0	0	0	0	0	0	0	0	0	0	0	0	0	0	0	0	0	0	0	0	0	0
B41	2	0	0	0	0	0	0	0	0	0	0	0	0	0	0	0	0	0	0	0	0	0	0	0	0	0	0	0	0	0	0	0	0	0	0	0	0	0	0	0
B27	2	0	0	0	0	0	0	0	0	0	0	0	0	0	0	0	0	0	0	0	0	0	0	0	0	0	0	0	0	0	0	0	0	0	0	0	0	0	0	0
B24	2	0	0	0	0	0	0	0	0	0	0	0	0	0	0	0	0	0	0	0	0	0	0	0	0	0	0	0	0	0	0	0	0	0	0	0	0	0	0	0
B39	1	0	0	0	0	0	0	0	0	0	0	0	0	0	0	0	0	0	0	0	0	0	0	0	0	0	0	0	0	0	0	0	0	0	0	0	0	0	0	0
B43	2	0	0	0	0	0	0	0	0	0	0	0	0	0	0	0	0	0	0	0	0	0	0	0	0	0	0	0	0	0	0	0	0	0	0	0	0	0	0	0
B20	2	0	0	0	0	0	0	0	0	0	0	0	0	0	0	0	0	0	0	0	0	0	0	0	0	0	0	0	0	0	0	0	0	0	0	0	0	0	0	0
B34	3	0	0	0	0	0	0	0	0	0	0	0	0	0	0	0	0	0	0	0	0	0	0	0	0	0	0	0	0	0	0	0	0	0	0	0	0	0	1	1
B27	1	0	0	0	0	0	0	0	0	0	0	0	0	0	0	0	0	0	0	0	0	0	0	0	0	0	0	0	0	0	0	0	0	0	0	0	0	0	1	1
B37	1	0	0	0	0	0	0	1	0	0	0	0	0	0	0	0	0	0	0	0	0	0	0	0	0	0	0	0	0	0	0	0	0	0	0	0	0	0	0	1
B25	2	0	0	0	0	0	0	0	0	0	0	0	0	0	0	0	0	1	0	0	0	0	0	0	0	0	0	0	0	0	0	0	0	0	0	0	0	0	0	1
B30	2	0	0	0	0	0	0	0	0	0	0	0	0	0	0	1	0	0	0	0	0	0	0	0	0	0	0	0	0	0	0	0	0	0	0	0	0	0	0	1
B21	1	0	0	0	0	0	0	0	0	0	0	0	0	0	0	0	0	0	0	0	0	0	0	0	0	0	0	0	0	0	0	0	0	0	1	0	0	0	0	1
B49	2	0	0	0	0	0	0	0	0	0	0	0	0	0	0	0	0	0	0	0	0	0	0	0	0	0	0	0	0	0	0	0	0	0	0	1	0	0	0	1
B40	2	0	0	0	0	0	0	0	0	0	0	0	0	0	0	0	0	0	0	0	0	0	0	0	0	0	0	0	1	1	0	0	0	0	0	0	0	0	0	2
BT8	2	0	0	0	0	0	0	0	1	0	0	0	0	0	0	0	0	0	0	0	0	0	0	0	0	0	0	0	0	0	0	0	0	0	0	0	1	0	0	2
BT17	2	0	0	0	0	0	0	0	0	0	0	0	0	0	0	0	0	0	0	0	0	0	0	0	0	0	0	0	0	0	0	0	0	0	0	0	1	0	1	2
BT21	1	0	0	0	0	0	0	0	0	0	0	0	0	0	0	0	0	0	0	0	0	0	0	0	0	0	0	0	0	0	0	1	0	1	0	0	0	0	0	2
BT5	1	0	0	0	0	0	0	0	0	0	0	0	0	0	0	0	0	0	1	0	0	0	0	0	0	0	0	0	0	0	0	1	0	0	0	0	0	1	0	3
B14	3	0	0	0	0	0	0	0	0	0	0	0	0	0	0	0	0	0	0	0	0	0	1	0	0	0	0	0	0	0	0	0	0	0	1	0	0	1	0	3
B31	1	0	0	0	0	0	0	0	0	0	0	0	0	0	0	1	0	0	0	0	0	0	0	0	0	0	0	0	0	0	0	0	0	0	0	1	0	1	0	3
B42	1	0	0	0	0	0	0	0	0	0	0	0	0	0	0	0	0	0	0	0	1	0	0	0	0	0	0	0	0	0	0	0	0	0	0	0	1	1	0	3
B44	2	0	0	0	0	0	0	0	0	0	0	0	0	0	0	0	0	0	0	0	0	0	0	0	0	0	0	0	0	1	1	1	0	0	0	0	0	0	0	3
B35	1	0	0	0	0	0	0	0	0	0	0	0	0	0	0	0	0	1	0	0	1	0	0	0	0	0	0	0	0	0	0	0	0	0	1	0	0	0	0	3
BT15	1	0	0	0	0	0	0	0	0	0	0	0	0	0	0	0	0	0	0	0	0	0	0	0	0	0	0	1	0	0	0	0	0	0	0	0	1	1	0	3
B15	3	1	0	0	0	0	0	0	0	0	0	0	0	0	0	0	0	0	0	0	1	0	0	0	0	0	0	0	0	0	0	0	0	0	0	0	1	1	0	4
B45	3	0	0	0	0	0	0	0	0	0	0	0	0	0	0	0	0	0	0	0	1	0	0	0	0	0	0	0	1	1	0	0	0	1	0	0	1	0	0	4

(continued)

Table 9.9 (continued)   Grave Goods in Burials in the Bannadi Cemetery in Northern Thailand

Burial	Type	Type of Article 1	2	3	4	5	6	7	8	9	10	11	12	13	14	15	16	17	18	19	20	21	22	23	24	25	26	27	28	29	30	31	32	33	34	35	36	37	38	Sum
B46	3	0	0	0	0	0	0	0	0	0	0	0	0	0	0	0	0	0	0	0	0	0	0	0	0	0	0	0	0	1	1	1	0	1	0	0	0	1	0	4
B17	1	0	0	0	0	0	0	0	0	0	0	0	0	0	0	0	0	1	0	0	0	0	0	0	0	0	0	0	0	0	1	0	1	1	0	0	1	0	0	4
B10	2	0	0	0	0	0	0	0	0	0	0	0	0	1	0	0	0	0	0	0	0	0	1	0	0	0	0	0	0	0	0	0	0	0	0	1	0	0	1	4
BT16	2	0	0	0	0	0	0	0	0	0	0	0	0	0	0	0	0	0	0	0	0	0	1	0	0	0	0	1	0	0	1	0	0	0	0	0	1	0	1	4
B26	2	0	0	0	0	0	0	0	0	0	0	0	0	0	0	0	0	0	0	0	0	0	0	0	0	0	0	0	0	0	0	0	0	0	1	1	1	0	1	4
B16	1	0	0	0	0	0	0	0	0	0	0	0	0	0	0	0	0	0	0	0	0	1	0	0	0	0	0	0	0	0	0	0	0	0	1	1	0	1	1	5
B29	3	0	0	0	1	0	0	0	0	0	0	0	1	0	0	0	0	0	0	0	0	0	0	0	0	0	0	0	0	0	0	0	0	0	1	0	0	1	0	5
B19	3	0	0	1	0	0	0	0	0	0	0	0	0	1	0	1	0	0	0	0	0	0	0	0	0	1	1	0	0	0	0	0	0	0	1	0	1	1	1	6
B32	2	0	0	0	0	0	0	0	0	0	0	0	0	0	0	0	0	0	0	0	0	0	0	0	1	0	0	0	0	1	0	1	1	1	0	1	0	0	1	6
B38	3	0	0	0	0	0	0	0	0	0	0	0	0	0	0	0	0	0	0	0	0	0	0	1	0	1	0	0	0	1	1	1	0	1	0	1	1	1	1	7
B36	2	0	0	0	0	0	0	0	0	1	0	1	0	0	0	0	0	0	0	0	0	0	0	0	1	0	0	1	0	0	1	0	0	1	1	1	1	1	1	7
B12	2	0	0	0	0	0	0	0	0	0	0	1	0	0	0	0	0	0	0	0	0	0	0	0	0	0	1	0	0	0	0	0	0	0	1	1	1	1	1	8
BT12	1	0	0	0	0	1	0	0	0	0	1	0	0	0	1	0	0	0	0	0	0	0	0	0	0	0	0	1	1	0	0	0	0	0	1	1	1	1	1	8
B47	1	1	0	0	1	0	0	0	0	0	0	0	0	0	0	0	0	0	1	1	1	0	0	0	0	0	0	0	1	1	0	1	0	0	0	0	1	0	1	9
B18	2	0	0	0	0	0	0	0	0	0	0	0	1	0	0	1	0	0	1	1	1	0	0	1	0	0	1	0	0	0	1	1	1	1	1	0	1	1	1	9
B48	2	0	0	0	0	1	1	0	0	1	1	0	0	1	1	1	1	1	1	1	1	1	0	0	1	1	1	0	0	0	1	1	1	1	1	1	1	1	1	11
Sum		1	1	1	1	1	1	0	0	1	1	1	1	1	1	1	1	1	1	0	1	1	2	2	3	3	3	3	4	6	6	6	7	8	9	12	15	16	18	144

Note: Body types: 1, adult male; 2, adult female; 3, child.

References

Digby, P.G.N. and Kempton, R.A. (1987), *Multivariate Analysis of Ecological Communities*, Chapman and Hall, London.

Everitt, B., Landau, S., and Leese, M. (2001), *Cluster Analysis*, 4th ed., Edward Arnold, London.

Hartigan, J. (1975), *Clustering Algorithms*, Wiley, New York.

Hintze, J. (2001), *NCSS and PASS*, Number Cruncher Statistical Systems, Kaysville, UT; available on-line at www.ncss.com.

Persson, S. (1981), Ecological indicator values as an aid in the interpretation of ordination diagrams, *J. Ecology*, 69, 71–84.

Romesburg, H.C. (1984), *Cluster Analysis for Researchers*, Lifetime Learning Publications, Belmont, CA.

chapter ten

Canonical correlation analysis

10.1 *Generalizing a multiple regression analysis*

In some sets of multivariate data, the variables divide naturally into two groups. A canonical correlation analysis can then be used to investigate the relationships between the two groups. A case in point is the data that are provided in Table 1.3. There 16 colonies of the butterfly *Euphydryas editha* in California and Oregon are considered. For each colony, values are available for four environmental variables and six gene frequencies. An obvious question to be considered is what relationships, if any, exist between the gene frequencies and the environmental variables. One way to investigate this is through a canonical correlation analysis.

Another example was provided by Hotelling (1936) in which he described a canonical correlation analysis for the first time. This example involved the results of tests for reading speed (X_1), reading power (X_2), arithmetic speed (Y_1), and arithmetic power (Y_2) for 140 seventh-grade schoolchildren. The specific question that was addressed was whether or not reading ability (as measured by X_1 and X_2) is related to arithmetic ability (as measured by Y_1 and Y_2).

The approach that a canonical correlation analysis takes to answering this question is to search for a linear combination of X_1 and X_2

$$U = a_1 X_1 + a_2 X_2$$

and a linear combination of Y_1 and Y_2

$$V = b_1 Y_1 + b_2 Y_2$$

where these are chosen to make the correlation between U and V as large as possible. This is somewhat similar to the idea behind a principal components analysis, except that here a correlation is maximized instead of a variance.

With X_1, X_2, Y_1, and Y_2 standardized to have unit variances, Hotelling found that the best choices for U and V with the reading and arithmetic example were

$$U = -2.78X_1 + 2.27X_2$$

and

$$V = -2.44Y_1 + 1.00Y_2$$

where these two variables have a correlation of 0.62. It can be seen that U measures the difference between reading power and speed, and V measures the difference between arithmetic power and speed. Hence, it appears that children with a large difference between X_1 and X_2 also tended to have a large difference between Y_1 and Y_2. It is this aspect of reading and arithmetic that shows the most correlation.

In a multiple regression analysis, a single variable Y is related to two or more variables X_1, X_2, ..., X_p to see how Y is related to the X variables. From this point of view, canonical correlation analysis is a generalization of multiple regression in which several Y variables are simultaneously related to several X variables.

In practice, more than one pair of canonical variables can be calculated from a set of data. If there are p variables X_1, X_2, ..., X_p and q variables Y_1, Y_2, ..., Y_q, then there can be up to the minimum of p and q pairs of variables. That is to say, linear relationships

$$U_1 = a_{11}X_1 + a_{12}X_2 + ... + a_{1p}X_p$$
$$U_2 = a_{21}X_1 + a_{22}X_2 + ... + a_{2p}X_p$$

$$\cdot$$
$$\cdot$$
$$\cdot$$

$$U_r = a_{r1}X_1 + a_{r2}X_2 + ... + a_{rp}X_p$$

and

$$V_1 = b_{11}Y_1 + b_{12}Y_2 + ... + b_{1q}Y_q$$
$$V_2 = b_{21}Y_1 + b_{22}Y_2 + ... + b_{2q}Y_q$$

$$\cdot$$
$$\cdot$$
$$\cdot$$

$$V_r = b_{r1}Y_1 + b_{r2}Y_2 + ... + b_{rq}Y_q$$

can be established, where r is the smaller of p and q. These relationships are chosen so that the correlation between U_1 and V_1 is a maximum; the correlation between U_2 and V_2 is a maximum, subject to these variables' being uncorrelated with U_1 and V_1; the correlation between U_3 and V_3 is a maximum, subject to these variables' being uncorrelated with U_1, V_1, U_2, V_2; and so on. Each of the pairs of canonical variables (U_1, V_1), (U_2, V_2), ..., (U_r, V_r) then represents an independent dimension in the relationship between the two sets of variables $(X_1, X_2, ..., X_p)$ and $(Y_1, Y_2, ..., Y_q)$. The first pair (U_1, V_1) has the highest possible correlation and is therefore the most important, the second pair (U_2, V_2) has the second highest correlation and is therefore the second most important, etc.

10.2   Procedure for a canonical correlation analysis

Assume that the $(p + q) \times (p + q)$ correlation matrix between the variables $X_1, X_2, ..., X_p$ and $Y_1, Y_2, ..., Y_q$ takes the following form when it is calculated from the sample for which the variables are recorded:

$$
\begin{array}{c}
\qquad\qquad X_1\ X_2...X_p\ \ Y_1\ Y_2...Y_q \\
\begin{array}{c}
X_1 \\ X_2 \\ \cdot \\ \cdot \\ \cdot \\ X_p \\ Y_1 \\ Y_2 \\ \cdot \\ \cdot \\ \cdot \\ Y_q
\end{array}
\left[
\begin{array}{c|c}
\begin{array}{c} p \times p \text{ matrix} \\ \mathbf{A} \end{array} & \begin{array}{c} p \times q \text{ matrix} \\ \mathbf{C} \end{array} \\
\hline
\begin{array}{c} q \times p \text{ matrix} \\ \mathbf{C}' \end{array} & \begin{array}{c} q \times q \text{ matrix} \\ \mathbf{B} \end{array}
\end{array}
\right]
\end{array}
$$

From this matrix, a $q \times q$ matrix $\mathbf{B}^{-1}\ \mathbf{C}'\ \mathbf{A}^{-1}\ \mathbf{C}$ can be calculated, and the eigenvalue problem

$$(\mathbf{B}^{-1}\ \mathbf{C}'\ \mathbf{A}^{-1}\ \mathbf{C} - \lambda\mathbf{I})\ \mathbf{b} = 0 \qquad\qquad (10.1)$$

can be considered. It turns out that the eigenvalues $\lambda_1 > \lambda_2 > ... > \lambda_r$ are then the squares of the correlations between the canonical variables, and the corresponding eigenvectors, $\mathbf{b}_1$, $\mathbf{b}_2$, ..., $\mathbf{b}_r$, give the coefficients of the Y variables for the canonical variables. Also, the coefficients of U_i, the ith canonical variable for the X variables, are given by the elements of the vector

$$\mathbf{a}_i = \mathbf{A}^{-1} \mathbf{C} \mathbf{b}_i \qquad (10.2)$$

In these calculations, it is assumed that the original X and Y variables are in a standardized form with means of zero and standard deviations of unity. The coefficients of the canonical variables are for these standardized variables.

From Equations 10.1 and 10.2, the ith pair of canonical variables are calculated as

$$U_i = \mathbf{a}'_i \mathbf{X}$$

and

$$V_i = \mathbf{b}'_i \mathbf{Y}$$

where

$$\mathbf{a}'_i = \left(a_{i1}, a_{i2}, \dots, a_{ip} \right)$$
$$\mathbf{b}'_i = \left(b_{i1}, b_{i2}, \dots, b_{iq} \right)$$
$$\mathbf{X}' = \left(x_1, x_2, \dots, x_p \right)$$
$$\mathbf{Y}' = \left(y_1, y_2, \dots, y_q \right)$$

with the X and Y values standardized. As they stand, U_i and V_i will have variances that depend upon the scaling adopted for the eigenvector $\mathbf{b}_i$. However, it is a simple matter to calculate the standard deviation of U_i for the data and divide the a_{ij} values by this standard deviation. This produces a scaled canonical variable U_i with unit variance. Similarly, if the b_{ij} values are divided by the standard deviation of V_i, then this produces a scaled V with unit variance.

This form of standardization of the canonical variables is not essential because the correlation between U_i and V_i is not affected by scaling. However, it may be useful when it comes to examining the numerical values of canonical variables for the individuals for which data are available.

10.3   Tests of significance

An approximate test for a relationship between the X variables as a whole and the Y variables as a whole was proposed by Bartlett (1947) for the situation where the data are a random sample from a multivariate normal distribution. This involves calculating the statistic

$$X^2 = -\{n - \tfrac{1}{2}(p+q+3)\} \sum_{i=1}^{r} \log_e (1-\lambda_i) \qquad (10.3)$$

where n is the number of cases for which data are available. The statistic can be compared with the percentage points of the chi-squared distribution with pq degrees of freedom (df), and a significantly large value provides evidence that at least one of the r canonical correlations is significant. A nonsignificant result indicates that even the largest canonical correlation can be accounted for by sampling variation only.

It is sometimes suggested that this test can be extended to allow the importance of each of the canonical correlations to be tested. Common suggestions are to:

1. Compare the ith contributions,

$$-\{n - \tfrac{1}{2}(p+q+3)\} \log_e (1-\lambda_i)$$

   to the right-hand side of Equation 10.3 with the percentage points of the chi-squared distribution having $p + q - 2i + 1$ df.
2. Compare the sum of the $(i + 1)$th to the rth contributions to the sum on the right-hand side of Equation 10.3 with the percentage points of the chi-squared distribution having $(p - i)(q - i)$ df.

Here, the first approach is assumed to be testing the ith canonical correlation directly, whereas the second is assumed to be testing for the significance of the $(i + 1)$th to rth canonical correlations as a whole.

The reason why these tests are not reliable is essentially the same as has already been discussed in Section 8.4 for a related test used with discriminant function analysis. This is that the ith largest sample canonical correlation may, in fact, have arisen from a population canonical correlation that is not the ith largest. Hence, the association between the r contributions to the right-hand side of Equation 10.3 and the r population canonical correlations is blurred. See Harris (1985, p. 211) for a further discussion about this matter.

There are also some modifications of the test statistic X^2 that are sometimes proposed to improve the chi-squared approximation for the distribution of this statistic when the null hypothesis holds and the sample size is small, but these will not be considered here.

10.4   *Interpreting canonical variates*

If

$$U_i = a_{i1}X_1 + a_{i2}X_2 + \dots + a_{ip}X_p$$

and

$$V_i = b_{i1}Y_1 + b_{i2}Y_2 + \dots + b_{iq}Y_q$$

then it seems that U_i can be interpreted in terms of the X variables with large coefficients a_{ij}, and V_i can be interpreted in terms of the Y variables with large coefficients b_{ij}. Of course, large here means large positive or large negative.

Unfortunately, correlations between the X and Y variables can upset this interpretation process. For example, it can happen that a_{i1} is positive, and yet the simple correlation between U_i and X_1 is negative. This apparent contradiction can come about when X_1 is highly correlated with one or more of the other X variables, with the result that part of the effect of X_1 is accounted for by the coefficients of these other X variables. In fact, if one of the X variables is almost a linear combination of the other X variables, then there will be an infinite variety of linear combinations of the X variables, some of them with very different a_{ij} values, that give virtually the same U_1 values. The same can be said about linear combinations of the Y variables.

The interpretation problems that arise with highly correlated X or Y variables should be familiar to users of multiple regression analysis. Exactly the same problems arise with the estimation of regression coefficients.

Actually, a fair comment seems to be that if the X or Y variables are highly correlated, then there can be no way of disentangling their contributions to canonical variables. However, people will undoubtedly continue to try to make interpretations under these circumstances.

Some authors have suggested that it is better to describe canonical variables by looking at their correlations with the X and Y variables rather than the coefficients a_{ij} and b_{ij}. For example, if U_i is highly positively correlated with X_1 then U_i can be considered to reflect X_1 to a large extent. Similarly, if V_i is highly negatively correlated with Y_1 then V_i can be considered to reflect the opposite of Y_1 to a large extent. This approach does at least have the merit of bringing out all of the variables to which the canonical variables seem to be related.

Example 10.1   *Environmental and genetic correlations for colonies of a butterfly*

The data in Table 1.3 can be used to illustrate the procedure for a canonical correlation analysis. Here there are 16 colonies of the butterfly *Euphydryas editha* in California and Oregon. These vary with respect to four environmental variables (altitude, annual precipitation, annual maximum temperature, and annual minimum temperature) and six genetic variables (percentages of six phosphoglucose-isomerase [Pgi] genes as determined by electrophoresis). Any significant relationships between the environmental and genetic variables are interesting because they may indicate the adaptation of *E. editha* to local environments.

For this canonical correlation analysis, the environmental variables have been treated as the X variables and the gene frequencies as the Y variables. However, all of the six gene frequencies shown in Table 1.3 have not been used because they add up to 100%, which allows different linear combinations of these variables to have the same correlation with a combination of the X variables. To see this, suppose that the first pair of canonical variables are U_1 and V_1, where

$$V_1 = b_{11}Y_1 + b_{12}Y_2 + \ldots + b_{16}Y_6$$

Then V_1 can be rewritten by replacing Y_1 by 100 minus the sum of the other variables to give

$$V_1 = 100b_{11} + (b_{12} - b_{11})Y_2 + \ldots + (b_{16} - b_{11})Y_6$$

This means that the correlation between U_1 and V_1 is the same as that between

$$(b_{12} - b_{11})Y_2 + \ldots + (b_{16} - b_{11})Y_6$$

and U_1, because the constant $100b_{11}$ in the second linear combination has no effect on the correlation. Thus two linear combinations of the Y variables, possibly with very different coefficients, can serve just as well for the canonical variable. In fact, it can be shown that an infinite number of different linear combinations of the Y variables will serve just as well, and the same holds true for linear combinations of standardized Y variables.

This problem is overcome by removing one of the gene frequencies from the analysis. In this case, the 1.30 gene frequency was omitted. The data were also further modified by combining the low frequencies for the 0.40- and 0.60-mobility genes. Thus the X variables being considered are X_1 = altitude, X_2 = annual precipitation, X_3 = annual maximum temperature, and X_4 = annual minimum temperature, while the Y variables are Y_1 = frequency of 0.40- and 0.60-mobility genes, Y_2 = frequency of 0.80-mobility genes, Y_3 = frequency of 1.00-mobility genes, and Y_4 = frequency of 1.16-mobility genes. It is the standardized values of the variables that have been analyzed so that for the remainder of this example, X_i and Y_i refer to the standardized X and Y variables.

The correlation matrix for the eight variables is shown in Table 10.1, partitioned into the submatrices **A, B, C,** and **C′,** as described in Section 10.2.

Table 10.1 Correlation Matrix for Variables Measured on Colonies of *Euphydryas editha*, Partitioned into A, B, C, and C′ Submatrices

	X_1	X_2	X_3	X_4	Y_1	Y_2	Y_3	Y_4
X_1	1.000	0.568	−0.828	−0.936	−0.201	−0.573	0.727	−0.458
X_2	0.568	1.000	−0.479	−0.705	−0.468	−0.550	0.699	−0.138
X_3	−0.828	−0.479	1.000	0.719	0.224	0.536	−0.717	0.438
X_4	−0.936	0.705	0.719	1.000	0.246	0.593	−0.759	0.412
				A	C			
				C′	B			
Y_1	−0.201	−0.468	0.224	0.246	1.000	0.638	−0.561	−0.584
Y_2	−0.573	−0.550	0.536	0.593	0.638	1.000	−0.824	−0.127
Y_3	0.727	0.699	−0.717	−0.759	−0.561	−0.824	1.000	−0.264
Y_4	−0.458	−0.138	0.438	0.412	−0.584	−0.127	−0.264	1.000

The eigenvalues obtained from Equation 10.1 are 0.7425, 0.2049, 0.1425, and 0.0069. Calculation of square roots gives the corresponding canonical correlations of 0.8617, 0.4527, 0.3775, and 0.0833, respectively, and the canonical variables are found to be as follows:

$$U_1 = -0.09X_1 - 0.29X_2 + 0.48X_3 + 0.29X_4$$
$$V_1 = +0.54Y_1 + 0.42Y_2 - 0.10Y_3 + 0.82Y_4$$
$$U_2 = +2.31X_1 - 0.73X_2 + 0.45X_3 + 1.27X_4$$
$$V_2 = -1.66Y_1 - 2.20Y_2 - 3.71Y_3 + 2.77Y_4$$
$$U_3 = +3.02X_1 + 1.33X_2 + 0.57X_3 + 3.58X_4$$
$$V_3 = -3.56Y_1 - 1.35Y_2 - 3.86Y_3 - 2.86Y_4$$
$$U_4 = +1.43X_1 + 0.26X_2 + 1.72X_3 - 0.03X_4$$
$$V_4 = +0.60Y_1 - 1.44Y_2 - 0.58Y_3 + 0.58Y_4$$

There are four canonical correlations because this is the minimum of the number of X variables and the number of Y variables (which both happen to be equal to four).

Although the canonical correlations are quite large, they are not significantly so, according to Bartlett's test, because of the small sample size. It is found that $X^2 = 18.34$ with 16 df; the probability of a value this large from the chi-squared distribution is about 0.30.

Laying aside the lack of significance, it is interesting to see what interpretation can be given to the first pair of canonical variables. From the equation for U_1, it can be seen that this is mainly a contrast between X_3 (maximum temperature) and X_4 (minimum temperature) on the one hand, and X_2 (precipitation) on the other. For V_1, there are moderate to large positive coefficients for Y_1 (0.40 and 0.60 mobility), Y_2 (0.80 mobility), and Y_4 (1.16 mobility), and a small negative coefficient for Y_3 (1.00 mobility). It appears that the 0.40-, 0.60-, 0.80-, and 1.16-mobility genes tend to be frequent in colonies with high temperatures and low precipitation.

The correlations between the environmental variables and U_1 are:

	Altitude	Precipitation	Maximum temperature	Minimum temperature
U_1	−0.92	−0.77	0.90	0.92

This suggests that U_1 is best interpreted as a measure of high temperatures and low altitude and precipitation. The correlations between V_1 and the gene frequencies are:

	Mobility 0.40/0.60	Mobility 0.80	Mobility 1.00	Mobility 1.16
V_1	0.38	0.74	−0.96	0.49

In this case, V_1 comes out clearly as indicating a lack of mobility-1.00 genes.

The interpretations of U_1 and V_1 are not the same when made on the basis of the coefficients of the canonical functions as they are on the basis of correlations. For U_1, the difference is not great and only concerns the status of altitude, but for V_1, the importance of the mobility-1.00 genes is very different. On the whole, the interpretations based on correlations seem best and correspond with what is seen in the data. For example, colony GL has the highest altitude, high precipitation, the lowest temperatures, and the highest frequency of 1.00-mobility genes. This compares with colony UO with a low altitude, low precipitation, high temperature, and the lowest frequency of mobility-1.00 genes. However, as mentioned in the previous section, there are real problems with interpreting canonical variables when the variables that they are constructed from have high correlations. Table 10.1 shows that this is indeed the case with this example.

Figure 10.1 shows a plot of the values of V_1 against the values of U_1. It is immediately clear that the colony labeled DP is somewhat unusual compared with the other colonies because the value of V_1 is not similar to that for other colonies with about the same values for U_1. From the interpretations given for U_1 and V_1, it would seem that the frequency of mobility-1.00 genes is unusually high for a colony with this environment. Inspection of the data in Table 1.3 shows that this is the case.

Example 10.2   Soil and vegetation variables in Belize

For an example with a larger data set, consider part of the data collected by Green (1973) for a study on the factors influencing the location of prehistoric Maya habitation sites in the Corozal district of Belize in Central America. Table 10.2 shows four soil variables and four vegetation variables recorded for 2.5×2.5-km squares. Canonical correlation analysis can be used to study the relationship between these two groups of variables.

The soil variables are X_1 = percent soil with constant lime enrichment, X_2 = percent meadow soil with calcium groundwater, X_3 = percent soil with

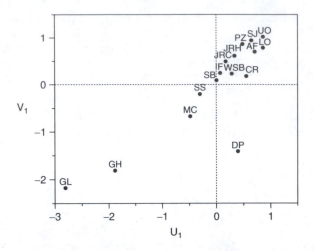

Figure 10.1 Plot of V_1 against U_1 for 16 colonies of *Euphydryas editha*.

coral bedrock under conditions of constant lime enrichment, and X_4 = percent alluvial and organic soils adjacent to rivers and saline organic soil at the coast. The vegetation variables are Y_1 = percent deciduous seasonal broadleaf forest; Y_2 = percent high and low marsh forest, herbaceous marsh, and swamp; Y_3 = percent cohune palm forest; and Y_4 = percent mixed forest. The percentages do not add to 100 for all of the squares, so there is no need to remove any variables before starting the analysis. It is the standardized values of these variables, with means of zero and standard deviations of one, that will be referred to for the rest of this example.

There are four canonical correlations (the minimum of the number of X variables and the number of Y variables), and they are found to be 0.762, 0.566, 0.243, and 0.122. The X^2 statistic of Equation 10.3 is found to equal 193.63 with 16 df, which is very highly significantly large when compared with the percentage points of the chi-squared distribution. Therefore, there is very strong evidence that the soil and vegetation variables are related. However, the original data are clearly not normally distributed, so this result should be treated with some reservations.

The canonical variables are found to be

$$U_1 = +1.34X_1 + 0.34X_2 + 1.13X_3 + 0.59X_4$$
$$V_1 = +1.71Y_1 + 1.07Y_2 + 0.22Y_3 + 0.52Y_4$$
$$U_2 = +0.41X_1 + 0.90X_2 + 0.23X_3 + 0.89X_4$$
$$V_2 = +0.64Y_1 + 1.47Y_2 + 0.27Y_3 + 0.28Y_4$$
$$U_3 = -0.44X_1 - 0.51X_2 + 0.18X_3 + 0.93X_4$$
$$V_3 = -0.18Y_1 - 0.24Y_2 + 0.93Y_3 + 0.22Y_4$$
$$U_4 = -0.44X_1 - 0.02X_2 + 0.72X_3 + 0.15X_4$$
$$V_4 = +0.12Y_1 + 0.01Y_2 + 0.26Y_3 - 0.93Y_4$$

Table 10.2 Soil and Vegetation Variables for 151 Squares, 2.5 × 2.5 km, in the Corozal Region of Belize

Square	X_1	X_2	X_3	X_4	Y_1	Y_2	Y_3	Y_4
1	40	30	0	30	0	25	0	0
2	20	0	0	10	10	90	0	0
3	5	0	0	50	20	50	0	0
4	30	0	0	30	0	60	0	0
5	40	20	0	20	0	95	0	0
6	60	0	0	5	0	100	0	0
7	90	0	0	10	0	100	0	0
8	100	0	0	0	20	80	0	0
9	0	0	0	10	40	60	0	0
10	15	0	0	20	25	10	0	0
11	20	0	0	10	5	50	0	0
12	0	0	0	50	5	60	0	0
13	10	0	0	30	30	60	0	0
14	40	0	0	20	50	10	0	0
15	10	0	0	40	80	20	0	0
16	60	0	0	0	100	0	0	0
17	45	0	0	0	5	60	0	0
18	100	0	0	0	100	0	0	0
19	20	0	0	0	20	0	0	0
20	0	0	0	60	0	50	0	0
21	0	0	0	80	0	75	0	0
22	0	0	0	50	0	50	0	0
23	30	10	0	60	0	100	0	0
24	0	0	0	50	0	50	0	0
25	50	20	0	30	0	100	0	0
26	5	15	0	80	0	100	0	0
27	60	40	0	0	10	90	0	0
28	60	40	0	0	50	50	0	0
29	94	5	0	0	90	10	0	0
30	80	0	0	20	0	100	0	0
31	50	50	0	0	25	75	0	0
32	10	40	50	0	75	25	0	0
33	12	12	75	0	10	90	0	0
34	50	50	0	0	15	85	0	0
35	50	40	10	0	80	20	0	0
36	0	0	100	0	100	0	0	0
37	0	0	100	0	100	0	0	0
38	70	30	0	0	50	50	0	0
39	40	40	20	0	50	50	0	0
40	0	0	100	0	100	0	0	0
41	25	25	50	0	100	0	0	0
42	40	40	0	20	80	20	0	0
43	90	0	0	10	100	0	0	0
44	100	0	0	0	100	0	0	0

(continued)

Table 10.2 (continued)   Soil and Vegetation Variables for 151 Squares, 2.5 × 2.5 km, in the Corozal Region of Belize

Square	X_1	X_2	X_3	X_4	Y_1	Y_2	Y_3	Y_4
45	100	0	0	0	90	10	0	0
46	10	0	0	90	100	0	0	0
47	80	0	0	20	100	0	0	0
48	60	0	0	30	80	0	0	0
49	40	0	0	0	0	30	0	0
50	50	0	0	50	100	0	0	0
51	50	0	0	0	40	0	0	0
52	30	30	0	20	30	60	0	0
53	20	20	0	40	0	100	0	0
54	20	80	0	0	0	100	0	0
55	0	10	0	60	0	75	0	0
56	0	50	0	30	0	75	0	0
57	50	50	0	0	30	70	0	0
58	0	0	0	60	0	60	0	0
59	20	20	0	60	0	100	0	0
60	90	10	0	0	70	30	0	0
61	100	0	0	0	100	0	0	0
62	15	15	0	30	0	40	0	0
63	100	0	0	0	25	75	0	0
64	95	0	0	5	90	10	0	0
65	95	0	0	5	90	10	0	0
66	60	40	0	0	50	50	0	0
67	30	60	10	10	50	10	0	0
68	50	0	50	50	100	0	0	0
69	60	30	0	10	60	40	0	0
70	90	8	0	2	80	20	0	0
71	30	30	30	40	60	40	0	0
72	33	33	33	33	75	25	0	0
73	20	10	0	40	0	100	0	0
74	50	0	0	50	40	60	0	0
75	75	12	0	12	50	50	0	0
76	75	0	0	25	40	60	0	0
77	30	0	0	50	0	100	0	0
78	50	10	0	30	5	95	0	0
79	100	0	0	0	60	40	0	0
80	50	0	0	50	20	80	0	0
81	10	0	0	90	0	100	0	0
82	30	30	0	20	0	85	0	0
83	20	20	0	20	0	75	0	0
84	90	0	0	0	50	25	0	0
85	30	0	0	0	30	5	0	0
86	20	30	0	50	20	80	0	0
87	50	30	0	10	50	50	0	0
88	80	0	0	0	70	10	0	0
89	80	0	0	0	50	0	0	0

(continued)

Table 10.2 (continued) Soil and Vegetation Variables for 151 Squares, 2.5 × 2.5 km, in the Corozal Region of Belize

Square	X_1	X_2	X_3	X_4	Y_1	Y_2	Y_3	Y_4
90	60	10	0	25	80	15	0	0
91	50	0	0	0	75	0	0	0
92	70	0	0	0	75	0	0	0
93	100	0	0	0	85	15	0	0
94	60	30	0	0	40	60	0	0
95	80	20	0	0	50	50	0	0
96	100	0	0	0	100	0	0	0
97	100	0	0	0	95	5	0	0
98	0	0	0	60	0	50	0	0
99	30	20	0	30	0	60	0	40
100	15	0	0	35	20	30	0	0
101	40	0	0	45	70	20	0	0
102	30	0	0	45	20	40	0	20
103	60	10	0	30	10	65	5	20
104	40	20	0	40	0	25	0	75
105	100	0	0	0	70	0	0	30
106	100	0	0	0	40	60	0	0
107	80	10	0	10	40	60	0	0
108	90	0	0	10	10	0	0	90
109	100	0	0	0	20	10	0	70
110	30	50	0	20	10	90	0	0
111	60	40	0	0	50	50	0	0
112	100	0	0	0	80	10	0	10
113	60	0	0	40	60	10	30	0
114	50	50	0	0	0	100	0	0
115	60	30	0	10	25	75	0	0
116	40	0	0	60	30	20	50	0
117	30	0	0	70	0	50	50	0
118	50	20	0	30	0	100	0	0
119	50	50	0	0	25	75	0	0
120	90	10	0	0	50	50	0	0
121	100	0	0	0	60	40	0	0
122	50	0	0	50	70	30	0	0
123	10	10	0	80	0	100	0	0
124	50	50	0	0	30	70	0	0
125	75	0	0	25	80	20	0	0
126	40	0	0	60	0	100	0	0
127	90	10	0	10	75	25	0	0
128	45	45	0	55	30	70	0	0
129	20	35	0	80	10	90	0	0
130	80	0	0	20	70	30	0	0
131	100	0	0	0	90	0	0	0
132	75	0	0	25	50	50	0	0
133	60	5	0	40	50	50	0	0
134	40	0	0	60	60	40	0	0

(continued)

Table 10.2 (continued) Soil and Vegetation Variables for 151 Squares, 2.5 × 2.5 km, in the Corozal Region of Belize

Square	X_1	X_2	X_3	X_4	Y_1	Y_2	Y_3	Y_4
135	60	0	0	40	70	15	0	0
136	90	10	0	10	75	25	0	0
137	50	0	5	0	30	20	0	0
138	70	0	30	0	70	30	0	0
139	60	0	40	0	100	0	0	0
140	50	0	0	0	50	0	0	0
141	30	0	50	0	60	40	0	0
142	5	0	95	0	80	20	0	0
143	10	0	90	0	70	30	0	0
144	50	0	0	0	15	30	0	0
145	20	0	80	0	50	50	0	0
146	0	0	100	0	90	10	0	0
147	0	0	100	0	75	25	0	0
148	90	0	10	0	60	30	10	0
149	0	0	100	0	80	10	10	0
150	0	0	100	0	60	40	0	0
151	0	40	60	40	50	50	0	0

Note: X_1 = % soil with constant lime enrichment, X_2 = % meadow soil with calcium ground-water, X_3 = % soil with coral bedrock under conditions of constant lime enrichment, and X_4 = % alluvial and organic soils adjacent to rivers and saline organic soil at the coast. Y_1 = % deciduous seasonal broadleaf forest; Y_2 = % high and low marsh forest, herbaceous marsh, and swamp; Y_3 = % cohune palm forest; and Y_4 = % mixed forest.

Table 10.3 Correlations between the Canonical Variables and the X and Y Variables

	U_1	U_2	U_3	U_4		V_1	V_2	V_3	V_4
X_1	0.55	−0.23	0.00	−0.80	Y_1	0.77	−0.58	−0.08	0.24
X_2	−0.02	0.73	−0.68	−0.04	Y_2	−0.36	0.91	−0.19	0.03
X_3	0.41	−0.24	−0.18	0.86	Y_3	0.03	0.13	0.95	0.28
X_4	−0.35	0.55	0.74	0.19	Y_4	0.11	−0.03	0.26	−0.96

In fact, the linear combinations given here for U_1, V_1, U_2, and V_2 are not the ones that were output by the program used to do the calculations, because the output linear combinations all had negative coefficients for the X and Y variables. A switch in sign is justified because the correlation between $-U_i$ and $-V_i$ is the same as that between U_i and V_i. Hence $-U_i$ and $-V_i$ will serve as well as U_i and V_i as the ith canonical variables. Note, however, that switching signs for U_1, V_1, U_2, and V_2 has changed the signs of the correlations between these canonical variables and the X and Y variables, as shown in Table 10.3.

By considering the correlations shown in Table 10.3 (particularly those outside the range from –0.5 to +0.5), it appears that the canonical variables can be described as mainly measuring:

U_1: the presence of soil types 1 (soil with constant lime enrichment) and 3 (soil with coral bedrock under conditions of constant lime enrichment)

V_1: the presence of vegetation type 1 (deciduous seasonal broadleaf forest)

U_2: the presence of soil types 2 (meadow soil with calcium groundwater) and 4 (alluvial and organic soils adjacent to rivers and saline organic soil at the coast)

V_2: the presence of vegetation type 2 (high and low marsh forest, herbaceous marsh and swamp) and the absence of vegetation type 1

U_3: the presence of soil type 4 and the absence of soil type 2

V_3: the presence of vegetation type 3 (cohune palm forest)

U_4: the presence of soil type 3 and the absence of soil type 1

V_4: the absence of vegetation type 4 (mixed forest)

It appears, therefore, that the most important relationships between the soil and vegetation variables, as described by the first two pairs of canonical variables, are:

1. The presence of soil types 1 and 3 and the absence of soil type 4 are associated with the presence of vegetation type 1.
2. The presence of soil types 2 and 4 is associated with the presence of vegetation type 2 and the absence of vegetation type 1.

It is instructive to examine a draftsman's plot of the canonical variables and the case numbers, as shown in Figure 10.2. The strong correlations between U_1 and V_1 and between U_2 and V_2 are apparent, as might be expected. Perhaps the most striking thing shown by the plots is the unusual distributions of V_3 and V_4. Most values are very similar, at about –0.2 for V_3 and about +0.2 for V_4. However, there are extreme values for some cases (observations) between 100 and 120. Inspection of the data in Table 10.2 shows that these extreme cases are for squares where vegetation types 3 and 4 were present, which makes perfect sense from the definition of V_3 and V_4.

Before leaving this example, it is appropriate to mention a potential problem that has not yet been addressed. This concerns the spatial correlation in the data for squares that are close in space, and particularly those that are adjacent. If such correlation exists so that, for example, neighboring squares tend to have the same soil and vegetation characteristics, then the data do not provide 151 independent observations. In effect, the data set will be equivalent to independent data from some smaller number of squares. The effect of this will appear mainly in the test for the significance of the canonical

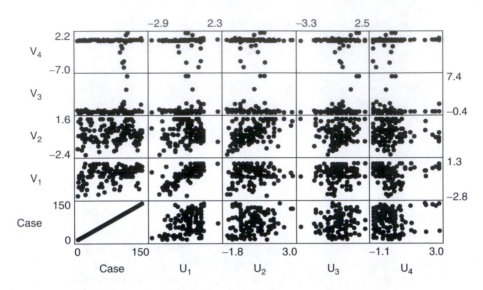

Figure 10.2 Draftsman's plot of canonical variables obtained from the data on soil and vegetation variables for 2.5-km squares in Belize. (Note that to enhance readability, some of the unit scales for the *x*- and *y*-axes appear, respectively, above and to the right of the plots.)

correlations as a whole, with a tendency for these correlations to appear to be more significant than they really are.

The same problem also potentially exists with the previous example on colonies of the butterfly *Euphydryas editha*, because some of the colonies were quite close in space. Indeed, it is a potential problem whenever observations are taken in different locations in space. The way to avoid the problem is to ensure that observations are taken sufficiently far apart that they are independent or close to independent, although this is often easier said than done. There are methods available to account for spatial correlation in data, but these are beyond the scope of this book.

10.5 Computer programs

The option for canonical correlation analysis is not as widely available in statistical packages as the options for multivariate analyses that were considered in earlier chapters. However, the larger packages certainly provide it, as shown in the Appendix to this book.

10.6 Further reading

There are not many books available that concentrate only on the theory and applications of canonical correlation analysis. Furthermore, the books that are available were written some time ago. A useful reference is the book by Giffins (1985) on applications of canonical correlation analysis in ecology. About half of this text is devoted to theory, and the remainder focuses on specific examples of plants. A shorter text with an emphasis on social sciences is by Thompson (1985).

10.7 Chapter summary

- Canonical correlation analysis can be thought of as a generalization of multiple regression that allows several dependent Y variables to be related to several explanatory X variables. Alternatively, it can be viewed as a method for understanding the relationships between any two groups of variables. It involves searching for linear combinations of X variables (U_1, U_2, etc.) that have the maximum possible correlations with linear combinations of Y variables (V_1, V_2, etc.).
- The process of calculating the U and V variables is an eigenvalue problem. These variables are called the canonical variates.
- Tests of significance for determining whether the correlations between the U and V variables are larger than expected by chance are described. However, there are some questions about the validity of these tests, particularly for small samples.
- The U and V variables can be interpreted in terms of the coefficients that they have for the X and Y variables, respectively, but this may lead to problems. Therefore, the correlations between the U and X variables and the correlations between the V and Y variables are often used instead for the interpretation process.
- Two examples of correlation are provided. The first concerns the relationship between gene frequencies of a butterfly at a number of colonies and the environmental conditions at those colonies. The second concerns the relationship between soil variables and vegetation variables on plots of land in the Corozal district of Belize in Central America.
- Computer programs for canonical correlation analysis are briefly discussed.
- Two older books devoted to the theory and applications of canonical correlation analysis are suggested as sources of further information.

Exercise

Table 10.4 shows the result of combining the data in Tables 1.5 and 6.7 on sources of protein and employment patterns for European countries for the 22 countries where these data coincide. Use canonical correlation analysis to investigate the relationship, if any, between the nature of the employment in a country and the type of food that is used for protein.

Table 10.4 Sources of Protein and Percentages Employed in Different Industry Groups for European Countries

Country	Sources of Protein [a]									Percentages Employed in Different Industry Groups [b]								
	RM	WM	EGG	MLK	FSH	CRL	SCH	PNO	F&G	AGR	MIN	MAN	PS	CON	SER	FIN	SPS	TC
Albania	10	1	1	9	0	42	1	6	2	55.5	19.4	0.0	0.0	3.4	3.3	15.3	0.0	3.0
Austria	9	14	4	20	2	28	4	1	4	7.4	0.3	26.9	1.2	8.5	19.1	6.7	23.3	6.4
Belgium	14	9	4	18	5	27	6	2	4	2.6	0.2	20.8	0.8	6.3	16.9	8.7	36.9	6.8
Bulgaria	8	6	2	8	1	57	1	4	4	19.0	0.0	35.0	0.0	6.7	9.4	1.5	20.9	7.5
Denmark	11	11	4	25	10	22	5	1	2	5.6	0.1	20.4	0.7	6.4	14.5	9.1	36.3	7.0
Finland	10	5	3	34	6	26	5	1	1	8.5	0.2	19.3	1.2	6.8	14.6	8.6	33.2	7.5
France	18	10	3	20	6	28	5	2	7	5.1	0.3	20.2	0.9	7.1	16.7	10.2	33.1	6.4
Greece	10	3	3	18	6	42	2	8	7	22.2	0.5	19.2	1.0	6.8	18.2	5.3	19.8	6.9
Hungary	5	12	3	10	0	40	4	5	4	15.3	28.9	0.0	0.0	6.4	13.3	0.0	27.3	8.8
Ireland	14	10	5	26	2	24	6	2	3	13.8	0.6	19.8	1.2	7.1	17.8	8.4	25.5	5.8
Italy	9	5	3	14	3	37	2	4	7	8.4	1.1	21.9	0.0	9.1	21.6	4.6	28.0	5.3
Netherlands	10	14	4	23	3	22	4	2	4	4.2	0.1	19.2	0.7	0.6	18.5	11.5	38.3	6.8
Norway	9	5	3	23	10	23	5	2	3	5.8	1.1	14.6	1.1	6.5	17.6	7.6	37.5	8.1
Poland	7	10	3	19	3	36	6	2	7	23.6	3.9	24.1	0.9	6.3	10.3	1.3	24.5	5.2
Portugal	6	4	1	5	14	27	6	5	8	11.5	0.5	23.6	0.7	8.2	19.8	6.3	24.6	4.8
Romania	6	6	2	11	1	50	3	5	3	22.0	2.6	37.9	2.0	5.8	6.9	0.6	15.3	6.8
Spain	7	3	3	9	7	29	6	6	7	9.9	0.5	21.1	0.6	9.5	20.1	5.9	26.7	5.8
Sweden	10	8	4	25	8	20	4	1	2	3.2	0.3	19.0	0.8	6.4	14.2	9.4	39.5	7.2
Switzerland	13	10	3	24	2	26	3	2	5	5.6	0.0	24.7	0.0	9.2	20.5	10.7	23.1	6.2
U.K.	17	6	5	21	4	24	5	3	3	2.2	0.7	21.3	1.2	7.0	20.2	12.4	28.4	6.5
USSR	9	5	2	17	3	44	6	3	3	18.5	0.0	28.8	0.0	10.2	7.9	0.6	25.6	8.4
Yugoslavia	4	5	1	10	1	56	3	6	3	5.0	2.2	38.7	2.2	8.1	13.8	3.1	19.1	7.8

[a] RM = red meat; WM = white meat; EGG = eggs; MLK = milk; FSH = fish; SCH = starchy foods; PNO = pulses, nuts, and oilseed; F&V = fruit and vegetables.

[b] AGR = agriculture forestry and fishing; MIN = mining and quarrying; MAN = manufacturing; PS = power and water supplies; CON = construction; SER = services; FIN = finance; SPS = social and personal services; TC = transport and communications.

References

Bartlett, M.S. (1947), The general canonical correlation distribution, *Ann. Mathematical Statistics*, 18, 1–17.

Giffins, R. (1985), *Canonical Analysis: a Review with Applications in Ecology*, Springer-Verlag, Berlin.

Green, E.L. (1973), Location analysis of prehistoric Maya sites in British Honduras, *Am. Antiquity*, 38, 279–293.

Harris, R.J. (1985), *A Primer of Multivariate Statistics*, Academic Press, Orlando.

Hotelling, H. (1936), Relations between two sets of variables, *Biometrika*, 28, 321–377.

Thompson, B. (1985), *Canonical Correlation Analysis: Uses and Interpretations*, Sage Publications, Thousand Oaks, CA.

chapter eleven

Multidimensional scaling

11.1   Constructing a map from a distance matrix

Multidimensional scaling is designed to construct a diagram showing the relationships between a number of objects, given only a table of distances between the objects. The diagram is thus a type of map that can be in one dimension (if the objects fall on a line), in two dimensions (if the objects lie on a plane), in three dimensions (if the objects can be represented by points in space), or in a higher number of dimensions (in which case a simple geometrical representation is not possible).

The fact that it may be possible to construct a map from a table of distances can be seen by considering the example of four objects — A, B, C, and D — shown in Figure 11.1. The distances between the objects are given in Table 11.1. For example, the distance from A to B, which is the same as the distance from B to A, is 6.0, while the distance of each objects to itself is always 0.0. It seems plausible that the map can be reconstructed from the array of distances. However, it is also apparent that a mirror image of the map, as shown in Figure 11.2 will have the same array of distances between objects. Consequently, it seems clear that a recovery of the original map will be subject to a possible reversal of this type.

It is also apparent that if more than three objects are involved, then they may not lie on a plane. In that case, the distance matrix will implicitly contain this information. For example, the distance array shown in Table 11.2 requires three dimensions to show the spatial relationships between the four objects. Unfortunately, with real data, it is not usually known how many dimensions are needed for a representation. Hence, with real data, a range of dimensions usually has to be tried.

The usefulness of multidimensional scaling comes from the fact that situations often arise where the underlying relationship between objects is not known, but a distance matrix can be estimated. For example, in psychology, subjects may be able to assess how similar or different individual pairs of objects are without being able to draw an overall picture of the relationships between the objects. Multidimensional scaling can then provide the picture.

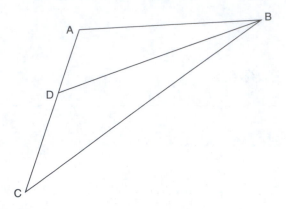

Figure 11.1 Four objects in two dimensions.

Table 11.1 Euclidean Distances between the Objects Shown in Figure 11.1

	A	B	C	D
A	0.0	6.0	6.0	2.5
B	6.0	0.0	9.5	7.8
C	6.0	9.5	0.0	3.5
D	2.5	7.8	3.5	0.0

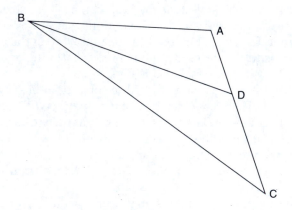

Figure 11.2 A mirror image of the objects in Figure 11.1 for which the distances between the objects are the same.

At the present time, there are a wide variety of data analysis techniques that go under the general heading of multidimensional scaling. Here, only the simplest will be considered, these being the classical methods proposed by Torgerson (1952) and Kruskal (1964a, 1964b). A related method called principal coordinates analysis is discussed in Chapter 12.

Table 11.2 A Matrix of Distances between Four Objects
in Three Dimensions

	A	B	C	D
A	0	1	$\sqrt{2}$	$\sqrt{2}$
B	1	0	1	1
C	$\sqrt{2}$	1	0	$\sqrt{2}$
D	$\sqrt{2}$	1	$\sqrt{2}$	0

11.2   Procedure for multidimensional scaling

A classical multidimensional scaling starts with a matrix of distances between n objects that has δ_{ij}, the distance from object i to object j, in the ith row and jth column. The number of dimensions for the mapping of objects is fixed for a particular solution at t (1 or more). Different computer programs use different methods for carrying out analysis, but generally something like the following steps are involved:

1. A starting configuration is set up for the n objects in t dimensions, i.e., coordinates $(x_1, x_2, \ldots, x_t)$ are assumed for each object in a t-dimensional space.
2. The Euclidean distances between the objects are calculated for the assumed configuration. Let d_{ij} be the distance between object i and object j for this configuration.
3. A regression of d_{ij} on δ_{ij} is made where, as mentioned above, δ_{ij} is the distance between object i and object j, according to the input data. The regression can be linear, polynomial, or monotonic. For example, a linear regression assumes that

$$d_{ij} = \alpha + \beta_{ij} + \varepsilon_{ij}$$

where ε_{ij} is an error term, while α and β are constants. A monotonic regression just assumes that if δ_{ij} increases, then d_{ij} either increases or remains constant, but no exact relationship between δ_{ij} and d_{ij} is assumed. The fitted distances obtained from the regression equation ($\hat{d}_{ij} = \alpha + \beta_{ij}$, assuming a linear regression) are called disparities. That is to say, the disparities $\hat{d}_{ij}$ are the data distances δ_{ij}, scaled to match the configuration distance d_{ij} as closely as possible.

4. The goodness of fit between the configuration distances and the disparities is measured by a suitable statistic. One possibility is Kruskal's stress formula 1, which is

$$\text{STRESS } 1 = \left\{ \sum \left(d_{ij} - \hat{d}_{ij} \right)^2 \Big/ \sum \hat{d}_{ij}^2 \right\}^{\frac{1}{2}} \tag{11.1}$$

The word *stress* is used here because the statistic is a measure of the extent to which the spatial configuration of points has to be stressed in order to obtain the data distances δ_{ij}.

5. The coordinates $(x_1, x_2, ..., x_t)$ of each object are changed slightly in such a way that the stress is reduced.

Steps 2 to 5 are repeated until it seems that the stress cannot be further reduced. The outcome of the analysis is then the coordinates of the n objects in t dimensions. These coordinates can be used to draw a map that shows how the objects are related. It is best when a good solution can be found in three or fewer dimensions, as a graphical representation of the n objects is then straightforward. Obviously this is not always possible.

Small values of STRESS 1 (close to 0) are desirable. However, defining what is meant by "small" for a good solution is not straightforward. As a rough guide, Kruskal and Wish (1978, p. 56) indicate that reducing the number of dimensions to the extent that STRESS 1 exceeds 0.1, or increasing the number of dimensions when STRESS 1 is already less than 0.05, is questionable. However, their discussion concerning choosing the number of dimensions involves more considerations than this. In practice, the choice of the number of dimensions is often made subjectively based on a compromise between the desire to keep the number small and the opposing desire to make the stress as small as possible. What is clear is that, in general, there is little point in increasing the number of dimensions if this only leads to a small decrease in the stress.

It is important to distinguish between metric multidimensional scaling and nonmetric dimensional scaling. In the metric case, the configuration distances d_{ij} and the data distances δ_{ij} are related by a linear or polynomial regression equation. With nonmetric scaling, all that is required is a monotonic regression, which means that only the ordering of the data distances is important. Generally, the greater flexibility of nonmetric scaling should make it possible to obtain a better low-dimensional representation of the data.

Example 11.1   Road distances between New Zealand towns

As an example of what can be achieved by multidimensional scaling, consider a map of the South Island of New Zealand that has been constructed from a table of the road distances between the 13 towns shown in Figure 11.3.

If road distances were proportional to geographic distances, it would be possible to recover the true map exactly by using a two-dimensional analysis. However, due to the absence of direct road links between many towns, road distances are in some cases far greater than geographic distances. Consequently, all that can be hoped for is a rather approximate recovery of the true map shown in Figure 11.3 from the road distances that are shown in Table 11.3.

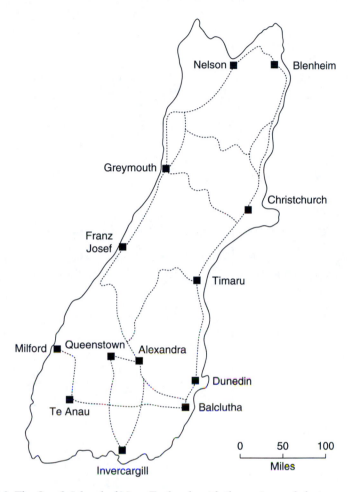

Figure 11.3 The South Island of New Zealand, with the main roads between 13 towns indicated by broken lines.

The computer program NCSS (Hintze, 2001) was used for the analysis. At step 3 of the procedure described above, a monotonic regression relationship was assumed between the map distances d_{ij} and the distances δ_{ij} given in Table 11.3. This gives what is sometimes called a classical nonmetric multidimensional scaling. The program produced a two-dimensional solution for the data using the algorithm described above. The final stress value was 0.041 as calculated using Equation 11.1.

The output from the program includes the coordinates of the 13 towns for the two dimensions produced in the analysis, as shown in Table 11.4. To maintain the north-south and east-west orientation that exists between the real towns, the signs of the values for the second dimension have been reversed to produce what is called new dimension 2. This sign reversal does not change the distances between the towns based on the two dimensions,

Table 11.3 Main Road Distances in Miles between 13 Towns in the South Island of New Zealand

	Alexandra	Balclutha	Blenheim	Christchurch	Dunedin	Franz Josef	Greymouth	Invercargill	Milford	Nelson	Queenstown	Te Anau	Timaru
Alexandra	—												
Balclutha	100	—											
Blenheim	485	478	—										
Christchurch	284	276	201	—									
Dunedin	126	50	427	226	—								
Franz Josef	233	493	327	247	354	—							
Greymouth	347	402	214	158	352	114	—						
Invercargill	138	89	567	365	139	380	493	—					
Milford	248	213	691	489	263	416	555	174	—				
Nelson	563	537	73	267	493	300	187	632	756	—			
Queenstown	56	156	494	305	192	228	341	118	178	572	—		
Te Anau	173	138	615	414	188	366	480	99	75	681	117	—	
Timaru	197	177	300	99	127	313	225	266	377	366	230	315	—

Table 11.4 Coordinates Produced by Multidimensional
Scaling Applied to the Distances between Towns in the South
Island of New Zealand

Town	Dimension [a]		
	1	2	New 2
Alexandra	0.11	0.07	−0.07
Balclutha	0.19	−0.08	0.08
Blenheim	−0.38	−0.16	0.16
Christchurch	−0.15	−0.11	0.11
Dunedin	0.13	−0.10	0.10
Franz Josef	−0.18	0.20	−0.20
Greymouth	−0.27	0.06	−0.06
Invercargill	0.26	−0.01	0.01
Milford	0.36	0.13	−0.13
Nelson	−0.45	−0.08	0.08
Queenstown	0.13	0.12	−0.12
Te Anau	0.28	0.08	−0.08
Timaru	−0.03	−0.13	0.13

[a] Dimension 2 is what was produced by the computer program used.
The signs of this axis have been reversed for the new dimension 2
to match the geographical locations of the real towns.

and the new dimension is therefore just as satisfactory as the original one.
If the sign is left unchanged, then the plot of the towns against the two
dimensions looks like a mirror image of the real map.

A plot of the towns using these coordinates is shown in Figure 11.4. A
comparison of this figure with Figure 11.3 indicates that the multidimensional scaling has been quite successful in recovering the real map. On the
whole, the towns are shown with the correct relationships to each other. An
exception is Milford. Because this town can be reached only by road through
Te Anau, the map produced by multidimensional scaling has made Milford
closest to Te Anau. In fact, Milford is geographically closer to Queenstown
than it is to Te Anau.

Example 11.2   The voting behavior of congressmen

For a second example of the value of multidimensional scaling, consider the
distance matrix shown in Table 11.5. Here the distances are between 15 New
Jersey congressmen in the U.S. House of Representatives. They are counts
of the number of voting disagreements on 19 bills concerned with environmental matters. For example, Congressmen Hunt and Sandman disagreed
8 out of the 19 times. Sandman and Howard disagreed 17 out of the 19 times,
etc. An agreement was considered to occur if two congressmen both voted
yes, both voted no, or both failed to vote. The table of distances was constructed from original data given by Romesburg (1984, p. 155).

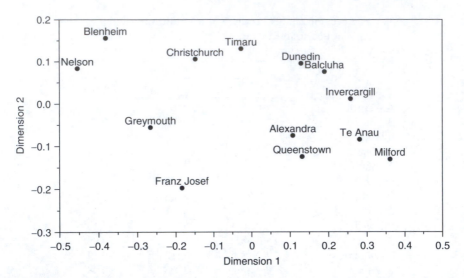

Figure 11.4 The map produced by a multidimensional scaling using the distances between New Zealand towns shown in Table 11.3.

Two analyses were carried out using the NCSS (Hintze, 2001) program. The first was a classical metric multidimensional scaling, which assumes that the distances of Table 11.5 are measured on a ratio scale. That is to say, it is assumed that doubling a distance value is equivalent to assuming that the configuration distance between two objects is doubled. This means that the regression at step 3 of the procedure described above is of the form

$$d_{ij} = \beta\delta_{ij} + \varepsilon_{ij}$$

where ε_{ij} is an error term and β is a constant. The stress values obtained for two-, three-, and four-dimensional solutions were found on this basis to be 0.237, 0.130, and 0.081, respectively.

The second analysis was carried out using a classical nonmetric scaling so that the regression of d_{ij} on δ_{ij} was assumed to be monotonic only. In this case, the stress values for two-, three-, and four-dimensional solutions were found to be 0.113, 0.066, and 0.044, respectively. The distinctly lower stress values for nonmetric scaling suggest that this is preferable to metric scaling for these data, and the three-dimensional nonmetric solution has only slightly more stress than the four-dimensional solution. This three-dimensional nonmetric solution is therefore the one that will be considered in greater detail. Table 11.6 shows the coordinates of the congressmen for the three-dimensional solution, and plots of the congressmen against the three dimensions are shown in Figure 11.5.

From Figure 11.5, it is clear that dimension 1 is largely reflecting party differences, because the Democrats fall on the left-hand side of the figure, and the Republicans, other than Rinaldo, fall on the right-hand side.

Table 11.5 The Distances between 15 Congressmen from New Jersey in the U.S. House of Representatives

	Hunt	Sandman	Howard	Thompson	Frelinghuysen	Forsythe	Widnall	Roe	Helstoski	Rodino	Minish	Rinaldo	Maraziti	Daniels	Pattern
Hunt (R)	0														
Sandman (R)	8	0													
Howard (D)	15	17	0												
Thompson (D)	15	12	9	0											
Frelinghuysen (R)	10	13	16	14	0										
Forsythe (R)	9	13	12	12	8	0									
Widnall (R)	7	12	15	13	9	7	0								
Roe (D)	15	16	5	10	13	12	17	0							
Helstoski (D)	16	17	5	8	14	11	16	4	0						
Rodino (D)	14	15	6	8	12	10	15	5	3	0					
Minish (D)	15	16	5	8	12	9	14	5	2	1	0				
Rinaldo (R)	16	17	4	6	12	10	15	3	1	2	1	0			
Maraziti (R)	7	13	11	15	10	6	10	12	13	11	12	12	0		
Daniels (D)	11	12	10	10	11	6	11	7	7	4	5	6	9	0	
Pattern (D)	13	16	7	7	11	10	13	6	5	6	5	4	13	9	0

Note: The numbers shown are the number of times that the congressmen voted differently on 19 environmental bills (R = Republican Party, D = Democratic Party).

Source: Romesburg, H.C. (1984). *Cluster Analysis for Researchers*, Lifetime Learning Publications, Belmont, CA.

Table 11.6 Coordinates of 15 Congressmen Obtained
from a Three-Dimensional Nonmetric
Multidimensional Scaling Based on Voting Behavior

Congressmen	Dimension		
	1	2	3
Hunt (R)	0.33	0.00	0.09
Sandman (R)	0.26	0.26	0.18
Howard (D)	−0.21	0.05	0.11
Thompson (D)	−0.12	0.22	−0.03
Frelinghuysen (R)	0.20	−0.06	−0.24
Forsythe (R)	0.13	−0.13	−0.06
Widnall (R)	0.33	0.00	−0.11
Roe (D)	−0.21	−0.05	0.09
Helstoski (D)	−0.22	0.02	−0.01
Rodino (D)	−0.16	−0.07	0.00
Minish (D)	−0.16	−0.03	−0.02
Rinaldo (R)	−0.18	0.01	−0.01
Maraziti (R)	0.19	−0.20	0.10
Daniels (D)	−0.02	−0.09	0.03
Pattern (D)	−0.16	0.05	−0.12

To interpret dimension 2, it is necessary to consider what it is about the voting of Sandman and Thompson, who have the highest two scores, that contrasts with Maraziti and Forsythe, who have the two lowest scores. This points to the number of abstentions from voting. Sandman abstained from nine votes and Thompson abstained from six votes, while individuals with low scores on dimension 2 voted all or most of the time.

Dimension 3 appears to have no simple or obvious interpretation, although it must reflect certain aspects of differences in voting patterns. It suffices to say that the analysis has produced a representation of the congressmen in three dimensions that indicates how they relate with regard to voting on environmental issues.

Figure 11.6 shows a plot of the distances between the congressmen for the original data (the disparities) against the points on the derived configuration. This indicates how well the three-dimensional model fits the data. A perfect representation of the data would show the data distances always increasing with the configuration distances. This is not obtained. Instead, there is a range of configuration distances associated with each of the discrete data distances. For example, data distances of 5 correspond to configuration distances from about 0.10 to about 0.16.

11.3 Computer programs

The calculations for the examples in this chapter were carried out using the NCSS computer package (Hintze, 2001). Only some of the standard statistical packages include a multidimensional scaling option. Some details are

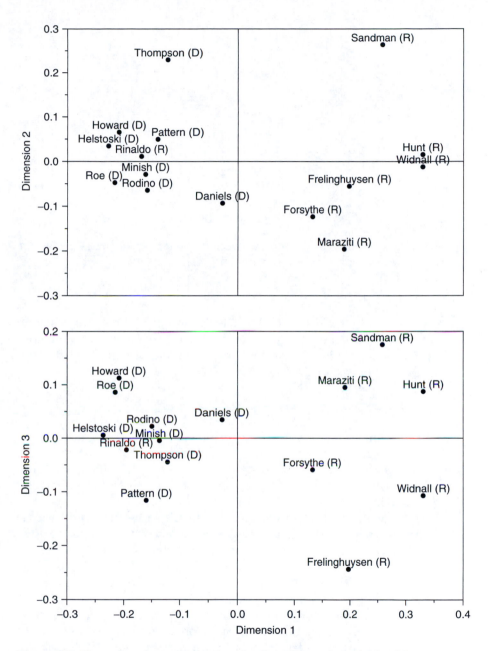

Figure 11.5 Plots of congressmen against the three dimensions obtained from a non-metric multidimensional scaling.

provided in the Appendix to this book. In general, it can be expected that different packages may use slightly different algorithms and therefore may not give exactly the same results. However, with good data, it can be hoped that the differences will not be substantial.

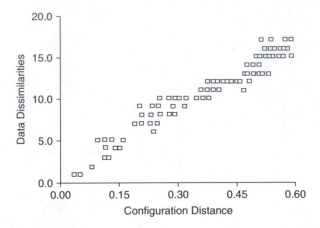

Figure 11.6 The original data distances between the congressmen plotted against the distances obtained for the fitted configuration.

11.4   Further reading

The classic book by Kruskal and Wish (1978) provides a short introduction to multidimensional scaling. More comprehensive treatments of the theory and applications of this topic and related topics are provided by Cox and Cox (1994) and Borg and Groenen (1997).

11.5   Chapter summary

- Multidimensional scaling is a technique designed to illustrate the relationship between a number of objects graphically, starting just with a table showing the distances between each of the objects; i.e., it is a method for producing a map showing how the objects are related.
- A basic algorithm with five steps is described for multidimensional scaling, with the differences between the data distances and the fitted distances (the disparities) measured by a statistic called the stress.
- The relationship between the data distances and the disparities can be assumed to be linear or a polynomial (with metric scaling), or just monotonic (with nonmetric scaling).
- The application of nonmetric scaling is illustrated by reconstructing a map of New Zealand towns, based on the road distances between them, and by considering the relationships between 15 New Jersey congressmen based on their voting behaviors.
- Computer programs for multidimensional scaling are briefly discussed.
- Suggestions are made for further reading.

Exercise

Consider the data in Table 1.5 about the percentages of people employed in different industries in 26 countries in Europe. From these data, construct a matrix of Euclidean distances between the countries using Equation 5.1. Carry out nonmetric multidimensional scaling using this matrix to find out how many dimensions are needed to represent the countries in a manner that reflects differences between their employment patterns.

References

Borg, I. and Groenen, P. (1997), *Modern Multidimensional Scaling: Theory and Applications*, Springer-Verlag, Berlin.

Cox, T.F. and Cox, M.A.A. (1994), *Multidimensional Scaling*, 2nd ed., Chapman and Hall/CRC, Boca Raton, FL.

Hintze, J. (2001), *NCSS and PASS*, Number Cruncher Statistical Systems, Kaysville, UT; available on-line at www.ncss.com.

Kruskal, J.B. (1964a), Multidimensional scaling by optimizing goodness of fit to a nonmetric hypothesis, *Psychometrics*, 29, 1–27.

Kruskal, J.B. (1964b), Nonmetric multidimensional scaling: a numerical method, *Psychometrics*, 29, 115–129.

Kruskal, J.B. and Wish, M. (1978), *Multidimensional Scaling*, Sage Publications, Thousand Oaks, CA.

Romesburg, H.C. (1984), *Cluster Analysis for Researchers*, Lifetime Learning Publications, Belmont, CA.

Torgerson, W.S. (1952), Multidimensional scaling, 1: theory and method, *Psychometrics*, 17, 401–419.

chapter twelve

Ordination

12.1 The ordination problem

The word *ordination* for a biologist means essentially the same as *scaling* does for a social scientist. Both words describe the process of producing a small number of variables that can be used to describe the relationship between a group of objects, starting either from a matrix of distances or similarities between the objects, or from the values of some variables measured on each object. From this point of view, many of the methods that have been described in earlier chapters can be used for ordination, and some of the examples have been concerned with this process. In particular, plotting female sparrows against the first two principal components of size measurements (Example 5.1), plotting European countries against the first two principal components for employment variables (Example 5.2), producing a map of the South Island of New Zealand from a table of distances between towns by multidimensional scaling (Example 11.1), and plotting New Jersey congressmen against axes obtained by multidimensional scaling based on voting behavior (Example 11.2) are all examples of ordination. In addition, discriminant function analysis can be thought of as a type of ordination that is designed to emphasize the differences between objects in different groups, while canonical correlation analysis can be thought of as a type of ordination that is designed to emphasize the relationships between two groups of variables measured on the same objects.

Although ordination can be considered to cover a diverse range of situations, in biology it is most often used as a means of summarizing the relationships between different species as determined from their abundances at a number of different locations or, alternatively, as a means of summarizing the relationships between different locations on the basis of the abundances of different species at those locations. It is this type of application that is considered particularly in the present chapter, although the examples involve archaeology as well as biology. The purpose of the chapter is to give more examples of the use of principal components analysis and multidimensional scaling in this context, and to describe the methods of principal coordinates analysis and correspondence analysis that have not been covered in earlier chapters.

12.2   Principal components analysis

Principal components analysis has already been discussed in Chapter 6. It may be recalled that it is a method whereby the values for variables X_1, X_2, ..., X_p, measured on each of n objects, are used to construct principal components Z_1, Z_2, ..., Z_p that are linear combinations of the X variables and are such that Z_1 has the maximum possible variance, Z_2 has the largest possible variance conditional on it being uncorrelated with Z_1, Z_3 has the maximum possible variance conditional on it being uncorrelated with both Z_1 and Z_2, and so on. The idea is that it may be possible, for some purposes, to replace the X variables with a smaller number of principal components, with little loss of information.

In terms of ordination, it can be hoped that the first two principal components are sufficient to describe the differences between the objects, because then a plot of Z_2 against Z_1 provides what is required. It is less satisfactory to find that three principal components are important, but a plot of Z_2 against Z_1 with values of Z_3 indicated may be acceptable. If four or more principal components are important, then, of course, a good ordination is not obtained, at least as far as a graphical representation is concerned.

Example 12.1   Plant species in the Steneryd Nature Reserve

Table 9.7 shows the abundances of 25 plant species on 17 plots from a grazed meadow in Steneryd Nature Reserve in Sweden, as described in Exercise 1 of Chapter 9, which was concerned with using the data for cluster analyses. Now it is an ordination of the plots that will be considered, and in this case, the variables for principal components analysis are the abundances of the plant species. In other words, in Table 9.7, the objects of interest are the plots (columns) and the variables are the species (rows).

Because there are more species than plots, the number of nonzero eigenvalues in the correlation matrix is determined by the number of plots. In fact, there are 16 nonzero eigenvalues, as shown in Table 12.1. The first three components account for about 69% of the variation in the data, which is not a particularly high amount. The coefficients for the first three principal components are shown in Table 12.2. They are all contrasts between the abundance of different species that may well be meaningful to a botanist, but no interpretations will be attempted here.

Figure 12.1 shows a draftsman's diagram of the plot number (1 to 17) and the first three principal components (PC). It is noticeable that the first component is closely related to the plot number. This reflects the fact that the plots are in the order of the abundance in the plots of species with a high response to light and a low response to moisture, soil reaction, and nitrogen. Hence the analysis has at least been able to detect this trend.

Table 12.1 Eigenvalues from a Principal Components Analysis of the Data in Table 9.7 Treating the Plots as the Objects of Interest and the Species Counts as the Variables

Component	Eigenvalue	% of Total	Cumulative %
1	8.79	35.17	35.17
2	5.59	22.34	57.51
3	2.96	11.82	69.33
4	1.93	7.72	77.04
5	1.58	6.32	83.37
6	1.13	4.52	87.89
7	0.99	3.97	91.86
8	0.55	2.18	94.04
9	0.40	1.60	95.64
10	0.35	1.40	97.04
11	0.20	0.78	97.82
12	0.18	0.70	98.53
13	0.13	0.51	99.04
14	0.12	0.46	99.50
15	0.07	0.30	99.80
16	0.05	0.20	100.00
Total	25.00	100.00	

Note: The values shown are for the coefficients of the standardized species abundances, with means of zero and standard deviations of one.

Example 12.2   Burials in Bannadi

For a second example of principal components ordination, the data shown in Table 9.8 concerning grave goods from a cemetery in Bannadi, in northeast Thailand, will be considered. The table (kindly supplied by Professor C.F.W. Higham) shows the presence or absence of 38 different types of article in each of 47 burials, with additional information on whether the body was of an adult male, adult female, or a child. In Exercise 2 of Chapter 9, it was suggested that cluster analysis be used to study the relationships between the burials. Now ordination is considered with the same end in mind. For a principal components analysis, the burials are the objects of interest, and the 38 types of grave goods provide the variables to be analyzed (presence or absence, i.e., 1 or 0, respectively). These variables were standardized before use so that the analysis was based on their correlation matrix.

In a situation like this, where only presence and absence data are available, it is common to find that a fairly large number of principal components are needed in order to account for most of the variation in the data. This is certainly the case here, with 11 components needed to account for 80% of the variance and 15 required to account for 90% of the variance. Obviously, there are far too many important principal components for a satisfactory ordination.

For this example, only the first four principal components will be considered, with the understanding that much of the variation in the original

Table 12.2 The First Three Principal Components for the Data in Table 9.7

Species	Z_1	Z_2	Z_3
Festuca ovina	0.30	0.01	−0.07
Anemone nemorosa	−0.25	0.02	−0.19
Stallaria holostea	−0.20	0.20	−0.19
Agrostis tenuis	0.17	0.14	0.01
Ranunculus ficaria	−0.11	−0.32	−0.07
Mercurialis perennis	−0.08	−0.31	0.02
Poa pratenis	−0.11	0.32	−0.11
Rumex acetosa	−0.01	0.34	0.23
Veronica chamaedrys	−0.15	0.36	−0.06
Dactylis glomerata	−0.23	0.15	0.18
Fraxinus excelsior (juv.)	−0.26	−0.11	0.17
Saxifraga granulata	0.13	0.24	0.23
Deschampsia flexuosa	−0.05	0.12	−0.45
Luzula campestris	0.28	0.09	0.00
Plantago lanceolata	0.27	0.11	0.26
Festuca rubra	−0.03	0.23	0.19
Hieracium pilosella	0.27	−0.02	0.05
Geum urbanum	−0.20	−0.18	0.29
Lathyrus montanus	−0.15	0.26	−0.19
Campanula persicifolia	−0.21	0.18	0.07
Viola riviniana	−0.24	0.17	0.11
Hepatica nobilis	−0.21	0.03	0.34
Achillea millefolium	0.29	0.03	0.10
Allium sp.	−0.18	−0.12	0.36
Trifolim repens	0.21	0.11	0.22

data is not accounted for. In fact, the four components correspond to eigenvalues of 5.29, 4.43, 3.65, and 3.34, while the total of all the eigenvalues is 38 (the number of types of articles). Thus these components account for 13.9%, 11.6%, 9.6%, and 8.8%, respectively, of the total variance, and between them account for 43.9% of the variance.

The coefficients of the standardized presence-absence variables are shown in Table 12.3, with the largest values (arbitrarily set at an absolute value greater than 0.2) underlined. To aid in interpretation, the signs of the coefficients have been reversed if necessary from what was given by the computer output in order to ensure that the values of all the components are positive for burial B48, which has the largest number of items present. This is allowable because switching the signs of all the coefficients for a component does not change the percentage of variation explained by the component; the direction of the signs is merely an accidental outcome of the numerical methods used to find the eigenvectors of the correlation matrix.

From the large coefficients of component 1, it can be seen that this is indicating the presence of articles type 18, 19, 20, 23, 25, 26, 32, 34, and 37, and the absence of articles type 3, 5, 6, 14, and 28. There is no grave with

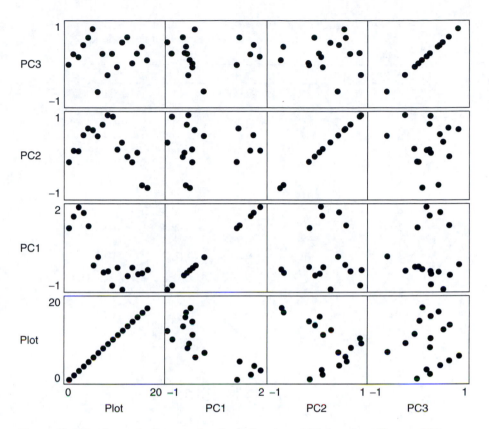

Figure 12.1 Draftsman's diagram for the ordination of 17 plots from Steneryd Nature Reserve.

exactly this composition, but the component measures the extent to which each of the graves matches this model. The other components can also be interpreted in a similar way from the coefficients in Table 12.3.

Figure 12.2 shows a draftsman's plot of the total number of goods, the type of body, and the first four principal components. From studying this, it is possible to draw some conclusions about the nature of the graves. For example, it seems that male graves tend to have low values and female graves to have high values for principal component 1, possibly reflecting a difference in grave goods associated with sex. Also, grave B47 has an unusual composition in comparison to the other graves. However, the fact that four principal components are being considered makes a simple interpretation of the results difficult.

12.3 Principal coordinates analysis

Principal coordinates analysis is similar to metric multidimensional scaling, which was discussed in Chapter 11. Both methods start with a matrix of

Table 12.3 Coefficients of Standardized Presence-Absence Data for the First Four Principal Components of the Bannadi Data

Article	PC1	PC2	PC3	PC4
1	0.01	−0.02	0.01	0.00
2	−0.09	−0.04	−0.02	0.52
3	−0.23	0.39	−0.01	−0.03
4	−0.09	−0.04	−0.02	0.52
5	−0.23	0.39	−0.01	−0.03
6	−0.23	0.39	−0.01	−0.03
7	−0.02	−0.05	−0.02	−0.02
8	−0.03	−0.02	−0.02	−0.02
9	0.17	0.12	0.33	0.06
10	0.15	0.09	0.04	0.07
11	0.05	0.03	0.03	−0.02
12	0.12	0.04	0.11	0.05
13	−0.01	−0.05	−0.02	−0.09
14	−0.23	0.39	−0.01	−0.03
15	0.00	0.00	0.03	−0.01
16	0.17	0.12	0.33	0.06
17	−0.01	−0.05	−0.02	−0.09
18	0.22	0.15	−0.38	0.04
19	0.22	0.15	−0.38	0.04
20	0.22	0.15	−0.38	0.04
21	−0.09	−0.04	−0.02	0.52
22	0.00	−0.04	0.01	0.01
23	0.27	0.17	−0.24	0.08
24	0.03	0.03	0.05	−0.07
25	0.26	0.15	0.28	0.11
26	0.26	0.15	0.28	0.11
27	0.08	0.02	0.02	0.04
28	−0.22	0.19	−0.02	0.26
29	−0.17	0.17	0.01	−0.08
30	0.17	0.11	0.00	−0.05
31	0.08	0.03	0.18	−0.04
32	0.27	0.14	0.04	0.03
33	−0.02	0.00	0.06	−0.12
34	0.23	0.09	−0.10	0.03
35	0.04	0.01	0.17	0.14
36	−0.07	0.15	0.17	−0.08
37	0.26	0.11	0.07	0.02
38	0.12	0.22	0.05	−0.05

similarities or distances between a number of objects and endeavor to find ordination axes. However, they differ in the numerical approach that is used. Principal coordinates analysis uses an eigenvalue approach that can be thought of as a generalization of principal components analysis. However, multidimensional scaling, at least as defined in this book, attempts instead

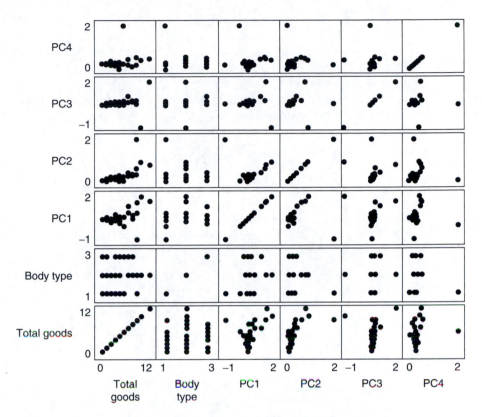

Figure 12.2 Draftsman's diagram for 47 Bannadi graves. The variables plotted are the total number of different types of goods, the type of remains (1 = adult male, 2 = adult female, 3 = child), and the first four principal components.

to minimize the stress, where this is a measure of the extent to which the positions of objects in a t-dimensional configuration fail to match the original distances or similarities after appropriate scaling.

    To see the connection between principal coordinates analysis and principal components analysis, it is necessary to recall some of the theoretical results concerning principal components analysis from Chapter 6, and to use some further results that are mentioned here for the first time. In particular:

1.   The ith principal component is a linear combination

$$Z_i = a_{i1}X_1 + a_{i2}X_2 + \ldots + a_{ip}X_p$$

    of the variables $X_1, X_2, \ldots, X_p$ that are measured on each of the objects being considered. There are p of these components, and the coefficients a_{ij} are given by the eigenvector $\mathbf{a}_i$ corresponding to the ith largest eigenvalue λ_i of the sample covariance matrix $\mathbf{C}$ of the X variables. That is to say, the equation

$$\mathbf{C a_i} = \lambda_i \, \mathbf{a_i} \tag{12.1}$$

is satisfied where $\mathbf{a_i'} = (a_{i1}, a_{i2}, \ldots, a_{ip})$. Also, the variance of Z_i is $\text{Var}(Z_i) = \lambda_i$, where this is zero if Z_i corresponds to a linear combination of the X variables that is constant.

2. If the X variables are coded to have zero means in the original data, then the $p \times p$ covariance matrix $\mathbf{C}$ has the form

$$\mathbf{C} = \begin{bmatrix} \sum x_{i1}^2 & \sum x_{i1} x_{i2} & \cdots & \sum x_{i1} x_{ip} \\ \sum x_{i2} x_{i1} & \sum x_{i2}^2 & \cdots & \sum x_{i2} x_{ip} \\ . & . & . & . \\ . & . & . & . \\ . & . & . & . \\ \sum x_{ip} x_{i1} & \sum x_{ip} x_{i2} & \cdots & \sum x_{ip}^2 \end{bmatrix} \Big/ (n-1)$$

where there are n objects, x_{ij} is the value of X_j for the ith object, and the summations are for i from 1 to n. Hence

$$\mathbf{C} = \mathbf{X'\, X}/(n-1) \tag{12.2}$$

where

$$\mathbf{X} = \begin{bmatrix} x_{i1} & x_{i2} & \cdots & x_{ip} \\ x_{21} & x_{22} & \cdots & x_{2p} \\ . & . & & . \\ . & . & & . \\ . & . & & . \\ x_{n1} & x_{n2} & \cdots & x_{np} \end{bmatrix}$$

is a matrix containing the original data values.

3. The symmetric $n \times n$ matrix

$$\mathbf{S} = \mathbf{XX'} = \begin{bmatrix} \sum x_{1j}^2 & \sum x_{1j} x_{2j} & \cdots & \sum x_{1j} x_{nj} \\ \sum x_{2j} x_{1j} & \sum x_{2j}^2 & \cdots & \sum x_{2j} x_{nj} \\ . & . & . & . \\ . & . & . & . \\ . & . & . & . \\ \sum x_{nj} x_{1j} & \sum x_{nj} x_{2j} & \cdots & \sum x_{nj}^2 \end{bmatrix} \tag{12.3}$$

where the summations for j from 1 to p can be thought of as containing measures of the similarities between the n objects being considered. This is not immediately apparent, but it is justified by considering the squared Euclidean distance from object i to object k, which is

$$d_{ik}^2 = \sum_{j=1}^{p} \left(x_{ij} - x_{kj}\right)^2$$

Expanding the right-hand side of this equation shows that

$$d_{ik}^2 = s_{ii} + s_{kk} - 2s_{ik} \tag{12.4}$$

where s_{ik} is the element in the ith row and kth column of $\mathbf{XX}'$. It follows that s_{ik} is a measure of the similarity between objects i and k because increasing s_{ik} means that the distance d_{ik} between the objects is decreased. Further, it is seen that s_{ik} takes the maximum value of $(s_{ii} + s_{kk})/2$ when $d_{ik} = 0$, which occurs when the objects i and k have identical values for the variables X_1 to X_p.

4. If the matrix

$$\mathbf{Z} = \begin{bmatrix} z_{i1} & z_{i2} & \cdots & z_{ip} \\ z_{21} & z_{22} & \cdots & z_{2p} \\ \cdot & \cdot & & \cdot \\ \cdot & \cdot & & \cdot \\ z_{n1} & z_{n2} & \cdots & z_{np} \end{bmatrix}$$

contains the values of the p principal components for the n objects being considered, then this can be written in terms of the data matrix **X** as

$$\mathbf{Z} = \mathbf{X}\,\mathbf{A}' \tag{12.5}$$

where the ith row of **A** is $\mathbf{a}_i'$, the ith eigenvector of the sample covariance matrix **C**. It is a property of **A** that $\mathbf{A}'\mathbf{A} = \mathbf{I}$; i.e., the transpose of **A** is the inverse of **A**. Thus postmultiplying both sides of Equation 12.5 by **A** gives

$$\mathbf{X} = \mathbf{Z}\,\mathbf{A} \tag{12.6}$$

This statement of results has been lengthy, but it has been necessary in order to explain principal coordinates analysis in relationship to principal components analysis. To see this relationship, note that from Equations 12.1 and 12.2

$$\mathbf{X}'\,\mathbf{X}\,\mathbf{a}_i / (n-1) = \lambda_i\,\mathbf{a}_i$$

Then premultiplying both sides of this equation by **X** and using Equation 12.3 gives

$$S(Xa_i) = (n - 1) \lambda_i (Xa_i)$$

or

$$Sz_i = (n - 1) \lambda_i z_i \tag{12.7}$$

where $z_i = Xa_i$ is a vector of length n, which contains the values of Z_i for the n objects being considered. Therefore, the ith largest eigenvalue of the similarity matrix $S = X'X$ is $(n - 1)\lambda_i$, and the corresponding eigenvector give the values of the ith principal component for the n objects.

Principal coordinates analysis consists of applying Equation 12.7 to an $n \times n$ matrix S of similarities between n objects that is calculated using any of the many available similarity indices. In this way, it is possible to find the principal components corresponding to S without necessarily measuring any variables on the objects of interest. The components will have the properties of principal components and, in particular, will be uncorrelated for the n objects.

Applying principal coordinates analysis to the matrix XX' will give essentially the same ordination as a principal components analysis on the data in X. The only difference will be in terms of the scaling given to the components. In principal components analysis, it is usual to scale the ith component to have the variance λ_i, but with a principal coordinates analysis, the component would usually be scaled to have a variance of $(n - 1)\lambda_i$. This difference is immaterial because it is only the relative values of objects on ordination axes that are important.

There are two complications that can arise in a principal coordinates analysis that must be mentioned. They occur when the similarity matrix being analyzed does not have all the properties of a matrix calculated from data using the equation $S = XX'$.

First, from Equation 12.3 it can be seen that the sums of the rows and columns of XX' are all zero. For example, the sum of the first row is

$$\sum x_{1j}^2 + \sum x_{1j} x_{2j} + \ldots + \sum x_{1j} x_{nj} = \sum x_{ij} \left(x_{1j} + x_{2j} + \ldots + x_{nj} \right)$$

where the summations are for j from 1 to p. This is zero because $x_{1j} + x_{2j} + \ldots + x_{nj}$ is n times the mean of X_j, and all the X variables are assumed to have zero means. Hence it is required that the similarity matrix S should have zero sums for rows and columns. If this is not the case, then the initial matrix can be double-centered by replacing the element s_{ik} in row i and column k by $s_{ik} - s_{i.} - s_{.k} + s_{..}$ where $s_{i.}$ is the mean of the ith row of S, $s_{.k}$ is the mean of the kth column of S, and $s_{..}$ is the mean of all the elements in S. The double-centered similarity matrix will have zero row and column means and is therefore more suitable for the analysis.

The second complication is that some of the eigenvalues of the similarity matrix may be negative. This is disturbing because the corresponding principal components appear to have negative variances! However, the truth is just that the similarity matrix could not have been obtained by calculating $S = XX'$ for any data matrix. With ordination, only the components associated with the largest eigenvalues are usually used, so that a few small negative eigenvalues can be regarded as being unimportant. Large negative eigenvalues suggest that the similarity matrix being used is not suitable for ordination.

Computer programs for principal coordinates analysis sometimes offer the option of starting with either a distance matrix or a similarity matrix. If a distance matrix is used, then it can be converted to a similarity matrix by transforming the distance d_{ik} to the similarity measure $s_{ik} = -d_{ik}^2/2$, as suggested by Equation 12.4.

Example 12.3   Plant species in the Steneryd Nature Reserve (revisited)

As an example of the use of principal coordinates analysis, the data considered in Example 12.1 on species abundances on plots in Steneryd Nature Reserve were reanalyzed using Manhattan distances between plots. That is, the distance between plots i and k was measured by $d_{ik} = \Sigma \,|\, x_{ij} - x_{kj}\,|$, where the summation is for j over the 25 species and x_{ij} denotes the abundance of species j on plot i as given in Table 9.7. Similarities were calculated as $s_{ik} = -d_{ik}^2/2$ and then double-centered before eigenvalues and eigenvectors were calculated.

The first two eigenvalues of the similarity matrix were found to be 97,638.6 and 55,659.5, which account for 47.3% and 27.0% of the sum of the eigenvalues, respectively. On the face of it, the first two components therefore give a good ordination, with 74.3% of the variation accounted for. The third eigenvalue is much smaller at 12,488.2 and accounts for only 6.1% of the total.

Figure 12.3 shows a draftsman's diagram of the plot number and the first two components. Both components show a relationship with the plot number which, as noted in Example 12.1, is itself related to the response of the different species to environmental variables. Actually, a comparison of this draftsman's diagram with the plots of Figure 12.1 shows that the first two axes from the principal coordinates analysis are really very similar to the first two principal components apart from a difference in scaling.

Example 12.4   Burials in Bannadi (revisited)

As an example of a principal coordinates analysis on presence-and-absence data, consider again the data in Table 9.8 on grave goods in the Bannadi cemetery in northeast Thailand. The analysis started with the matrix of unstandardized Euclidean distances between the 47 burials so that the distance from grave i to grave k was taken to be $d_{ik} = \sqrt{\{\Sigma(x_{ij} - x_{ki})^2\}}$, where the summation is for j from 1 to 38, and x_{ij} is 1 if the jth type of article is present in the ith burial, or is otherwise 0. A similarity matrix was then obtained, as

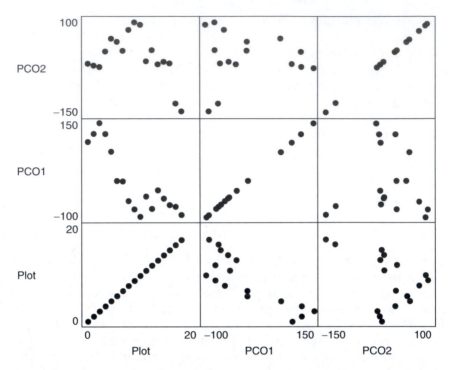

Figure 12.3 Draftsman's diagram for the ordination of 17 plots from Steneryd Nature Reserve based on a principal coordinate analysis on Manhattan distances between plots. The three variables are the plot number and the first two components (PCO1 and PCO2).

described in Example 12.3, and double-centered before eigenvalues and eigenvectors were obtained.

The principal coordinates analysis carried out in this manner gives the same result as a principal components analysis using unstandardized values for the X variables (i.e., carrying out a principal components analysis using the sample covariance matrix instead of the sample correlation matrix). The only difference in the results is in the scalings that are usually given to the ordination variables by principal components analysis and principal coordinates analysis.

The first four eigenvalues of the similarity matrix were 24.9, 19.3, 10.0, and 8.8, corresponding to 21.5%, 16.6%, 8.7%, and 7.6%, respectively, of the sum of all the eigenvalues. These components account for a mere 54.5% of the total variation in the data, but this is better than the 43.9% accounted for by the first four principal components obtained from the standardized data (Example 12.2).

Figure 12.4 shows a draftsman's diagram for the total number of goods in the burials, the type of remains (adult male, adult female, or child), and the first four components. The signs of the first and fourth components were switched from those shown on the computer output so as to make them

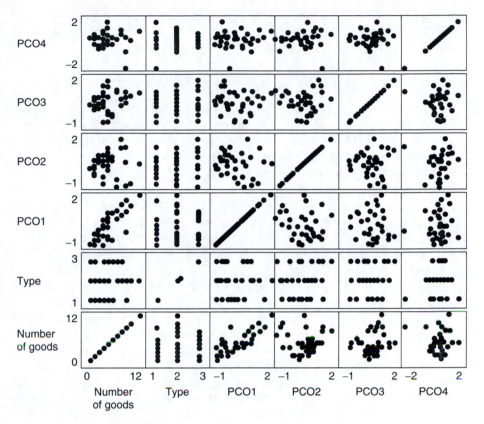

Figure 12.4 Draftsman's diagram for the 47 Bannadi graves. The six variables are the total number of different types of goods in a burial, an indicator of the type of remains (1 = adult male, 2 = adult female, 3 = child), and the first four components from a principal coordinates analysis (PCO1 to PCO4).

have positive values for burial B48, which contained the largest number of different types of grave goods. It can be seen from the diagram that the first component represents total abundance quite closely, but the other components are not related to this variable. Apart from this, the only obvious thing to notice is that one of the burials had a very low value for the fourth component. This is burial B47, which contained eight different types of article, of which four types were not seen in any other burial.

12.4 Multidimensional scaling

Multidimensional scaling has been discussed already in Chapter 11, where this is defined to be an iterative process for finding coordinates for objects on axes, in a specified number of dimensions, such that the distances between the objects match as closely as possible the distances or similarities that are provided in an input data matrix (Section 11.2). The method will not be discussed further in the present chapter except as required to present the results of using it on the two example sets of data that have been considered with the other methods of ordination.

Example 12.5   *Plant species in the Steneryd Nature Reserve (again)*

A multidimensional scaling of the 17 plots for the data in Table 9.7 was carried out using the computer program NMDS provided by Ludwig and Reynolds (1988). This performs a classical nonmetric type of analysis on a distance matrix, so that the relationship between the data distances and the ordination (configuration) distances is assumed to be only monotonic. A feature of the program is that after a solution is obtained, the axes are transformed to principal components. This ensures that the first axis accounts for the maximum possible variance in the ordination scores, the second axis accounts for the maximum possible remaining variance, and so on. The scores for the different axes are also made uncorrelated by this process.

For the example being considered, unstandardized Euclidean distances between the plots were used as input to the program. The stress values corresponding to solutions in from one to five dimensions were found to be 0.436, 0.081, 0.060, 0.023, and 0.021, so that a four-dimensional solution seems quite reasonable. Figure 12.5 shows a draftsman's diagram of the values for the plot numbers and the positions on these axes after they have been transformed to principal components. Comparison with Figures 12.1 and 12.3 shows that the first multidimensional scaling axis corresponds closely with the first principal component and the first principal coordinate axis, while the second multidimensional scaling axis, after a reversal in sign, corresponds closely with the second principal component and the second principal coordinate axis.

Example 12.6   *Burials in Bannadi (again)*

The same analysis as used in the last example was also applied to the data on burials at Bannadi shown in Table 9.8. Unstandardized Euclidean distances between the 47 burials were calculated using the presence-absence data (i.e., 1 or 0, respectively) in the table as values for 38 variables, and these distances provided the data for Ludwig and Reynolds's (1988) computer program NMDS. The stress levels obtained for solutions in one to five dimensions were 0.405, 0.221, 0.113, 0.084, and 0.060. Hence the three-dimensional solution seems reasonable, although the stress of 0.113 is quite large.

A draftsman's diagram for the three-dimensional solution is shown in Figure 12.6, with the axes reversed where necessary to ensure that a positive value is obtained for burial B48, which has the highest number of different types of goods. Comparison with Figure 12.2 shows that the first axis has a strong resemblance to the first principal component, but otherwise the relationship with ordinations from other methods is not immediately clear.

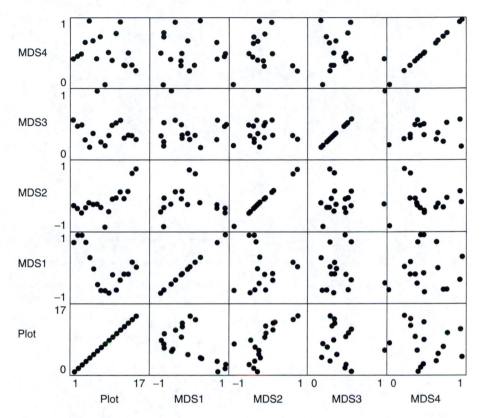

Figure 12.5 Draftsman's diagram for the ordination of 17 plots from Steneryd Nature Reserve based on nonmetric multidimensional scaling on Euclidean distances between the plots. The variables are the plot number and the coordinates for four axes (MDS1 to MDS4).

12.5 Correspondence analysis

Correspondence analysis as a method of ordination originated in the work of Hirschfeld (1935), Fisher (1940), and a school of French statisticians (Benzecri, 1992). It is now the most popular method of ordination for plant ecologists and is being used increasingly in other areas as well.

The method will be explained here in the context of the ordination of p sites on the basis of the abundance of n species, although it can be used equally well on data that can be presented as a two-way table of measures of abundance, with the rows corresponding to one type of classification and the columns to a second type of classification.

With sites and species, the situation is as shown in Table 12.4. Here there are a set of species values $a_1, a_2, ..., a_n$ associated with the rows of the table, and a set of site values $b_1, b_2, ..., b_p$ associated with the columns of the table. One interpretation of correspondence analysis is then that it is concerned with choosing species and site values so that they are as highly correlated

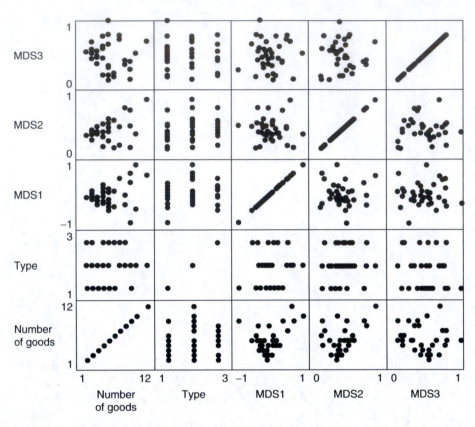

Figure 12.6 Draftsman's diagram for the 47 Bannadi graves. The variables plotted are the total number of different types of goods in a burial, an indicator of the type of remains (1 = adult male, 2 = adult female, 3 = child), and three axes from nonmetric multidimensional scaling using unstandardized Euclidean distances between the graves (MSD1 to MSD3).

Table 12.4 The Abundances (x) of n Species at p Sites, with the Species Values (a) and the Site Values (b)

Species	Site 1	Site 2	...	Site p	Row Sum	Species Value
1	x_{11}	x_{12}	...	x_{1p}	R_1	a_1
2	x_{21}	x_{22}	...	x_{2p}	R_2	a_2
.	.	.	...	.	.	.
.	.	.	...	.	.	.
.	.	.	...	.	.	.
n	x_{n1}	x_{n2}	...	x_{np}	R_n	a_n
Column sum	C_1	C_2	...	C_p		
Site value	b_1	b_2	...	b_p		

as possible for the bivariate distribution that is represented by the abundances in the body of the table. That is to say, the site and species values are chosen to maximize their correlation for the distribution where the number of times that species i occurs at site j is proportional to the observed abundance x_{ij}.

It turns out that the solution to this maximization problem is given by the set of equations

$$a_1 = \{(x_{11}/R_1)b_1 + (x_{12}/R_1)b_2 + \ldots + (x_{1p}/R_1)b_p\}/r$$
$$a_2 = \{(x_{21}/R_2)b_1 + (x_{22}/R_2)b_2 + \ldots + (x_{2p}/R_2)b_p\}/r$$

$$\cdot$$
$$\cdot$$

$$a_n = \{(x_{n1}/R_n)b_1 + (x_{n2}/R_n)b_2 + \ldots + (x_{np}/R_n)b_p\}/r$$

and

$$b_1 = \{(x_{11}/C_1)a_1 + (x_{21}/C_1)a_2 + \ldots + (x_{n1}/C_1)a_n\}/r$$
$$b_2 = \{(x_{12}/C_2)a_1 + (x_{22}/C_2)a_2 + \ldots + (x_{n2}/C_2)a_n\}/r$$

$$\cdot$$
$$\cdot$$

$$b_p = \{(x_{1p}/C_p)a_1 + (x_{2p}/C_p)a_2 + \ldots + (x_{np}/C_p)a_n\}/r$$

where R_i denotes the total abundance of species i, C_j denotes the total abundance at site j, and r is the maximum correlation being sought. Thus the ith species value a_1 is a weighted average of the site values, with site j having a weight that is proportional to x_{ij}/R_i, and the jth site value b_j is a weighted average of the species values, with species i having a weight that is proportional to x_{ji}/C_j.

The name "reciprocal averaging" is sometimes used to describe the equations just stated because the species values are (weighted) averages of the site values, and the site values are (weighted) averages of the species values. These equations are themselves often used as the starting point for justifying correspondence analysis as a means of producing species values as a function of site values, and vice versa. It turns out that the equations can be solved iteratively after they have been modified to remove the trivial solution with $a_i = 1$ for all i, $b_j = 1$ for all j, and $r = 1$. However, it is more instructive to write the equations in matrix form in order to solve them because that shows that there may be several possible solutions to the equations and that these can be found from an eigenvalue analysis.

In matrix form, the equations shown above become

$$\mathbf{a} = \mathbf{R}^{-1}\mathbf{X}\,\mathbf{b}/r \qquad\qquad (12.8)$$

and

$$\mathbf{b} = \mathbf{C}^{-1}\mathbf{X}'\,\mathbf{a}/r \qquad\qquad (12.9)$$

where $\mathbf{a}' = (a_1, a_2, \ldots, a_n)$ $\mathbf{b}' = (b_1\, b_2, \ldots, b_p)$, $\mathbf{R}$ is an $n \times n$ diagonal matrix with R_i in the ith row and ith column, $\mathbf{C}$ is a $p \times p$ diagonal matrix with C_j in the jth row and jth column, and $\mathbf{X}$ is an $n \times p$ matrix with x_{ij}, in the ith row and jth column.

If Equation 12.9 is substituted into Equation 12.8, then after some matrix algebra, it is found that

$$r^2\left(\mathbf{R}^{\frac{1}{2}}\mathbf{a}\right) = \left(\mathbf{R}^{\frac{1}{2}}\mathbf{X}\mathbf{C}^{-\frac{1}{2}}\right)\left(\mathbf{R}^{-\frac{1}{2}}\mathbf{X}\mathbf{C}^{-\frac{1}{2}}\right)'\left(\mathbf{R}^{\frac{1}{2}}\mathbf{a}\right) \qquad (12.10)$$

where $\mathbf{R}^{\frac{1}{2}}$ is a diagonal matrix with $\sqrt{R_i}$ in the ith row and ith column, and $\mathbf{C}^{\frac{1}{2}}$ is a diagonal matrix with $\sqrt{C_j}$ in the jth row and jth column. This shows that the solutions to the problem of maximizing the correlation are given by the eigenvalues of the $n \times n$ matrix

$$\left(\mathbf{R}^{-\frac{1}{2}}\mathbf{X}\mathbf{C}^{-\frac{1}{2}}\right)\left(\mathbf{R}^{-\frac{1}{2}}\mathbf{X}\mathbf{C}^{-\frac{1}{2}}\right)'$$

For any eigenvalue λ_k, the correlation between the species and site scores will be $r_k = \sqrt{\lambda_k}$, and the eigenvector for this correlation will be

$$\mathbf{R}^{\frac{1}{2}}\mathbf{a}_k = \left(\sqrt{R_1}a_{1k}, \sqrt{R_2}a_{2k}, \ldots, \sqrt{R_{na}}a_{nk}\right)'$$

where a_{ik} are the species values. The corresponding site values can be obtained from Equation 12.9 as

$$\mathbf{b}_k = \mathbf{C}^{-1}\mathbf{X}'\mathbf{a}_k/r_k$$

The largest eigenvalue will always be $r^2 = 1$, giving the trivial solution $a_i = 1$ for all i and $b_j = 1$ for all j. The remaining eigenvalues will be positive or zero and reflect different possible dimensions for representing the relationships between species and sites. These dimensions can be shown to be orthogonal, in the sense that the species and site values for one dimension will be uncorrelated with the species and site values in other dimensions for the data distribution of abundances x_{ij}.

Ordination by correspondence analysis involves using the species and site values for the first few largest eigenvalues that are less than 1, because these are the solutions for which the correlations between species values and site values are strongest. It is common to plot both the species and the sites on the same axes because, as noted earlier, the species values are an average of the site values and vice versa. In other words, correspondence analysis gives an ordination of both species and sites at the same time.

It is apparent from Equation 12.10 that correspondence analysis cannot be used on data that include a zero row sum because then the diagonal matrix $\mathbf{R}^{\frac{1}{2}}$ will have an infinite element. By a similar argument, zero column sums are not allowed either. This means that the method cannot be used on the burial data in Table 9.8, since some graves did not contain any articles. However, correspondence analysis can be used with presence-and-absence data when this problem is not present.

Example 12.7 Plant species in the Steneryd Nature Reserve (yet again)

Correspondence analysis was applied to the data for species abundances in the Steneryd Nature Reserve (Table 9.7). There were 16 eigenvalues less than 1 and the values are as follows, with their square roots (the correlations between species values and plot values) in parentheses: 0.665 (0.82), 0.406 (0.64), 0.199 (0.45), 0.136 (0.37), 0.094 (0.31), 0.074 (0.27), 0.057 (0.24), 0.028 (0.17), 0.020 (0.14), 0.019 (0.14), 0.010 (0.10), 0.008 (0.09), 0.007 (0.08), 0.005 (0.07), 0.003 (0.05), 0.001 (0.03). The first two or three might be considered to be important, but here only the species and plot values for the first two eigenvalues will be used for ordination.

Figure 12.7 shows a graph of the species and plot values for the eigenvalue of 0.406 (CORR2) against the species and plot values for the eigenvalue of 0.665 (CORR 1). Abbreviated names are shown for the species, and S1 to S17 indicate the sites. The ordination of sites is quite clear, with an almost perfect sequence from S1 on the right to S17 on the left, moving round the very distinct arch. The species are interspersed among the sites along the same arch from Mer-p (*Mercurialis perennis*) on the left to Hie-p (*Hieracium pilosella*) on the right. A comparison of the figure with Table 12.1 shows that it makes a good deal of sense. For example, *M. perennis* is abundant only on the highest numbered sites, and *H. pilosella* is abundant only on the lowest numbered sites.

The arch or horseshoe that appears in the ordination for this example is a common feature in the results of correspondence analysis, and it is also sometimes apparent with other methods of ordination as well. There is sometimes concern that this effect will obscure the nature of the ordination axes, and therefore some attention has been devoted to the development of ways to modify analyses to remove the effect, which is considered to be an artifact of the ordination method. With correspondence analysis, a method of detrending is usually used, and the resulting ordination method is then called detrended correspondence analysis (Hill and Gauch, 1980). Adjustments for other methods of ordination exist as well but seem to receive little use.

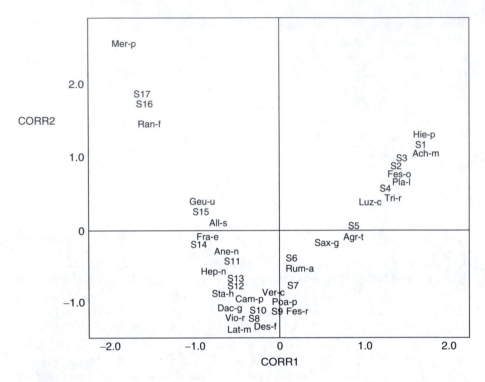

Figure 12.7 Plot of species and sites against the first two axes (CORR1 and CORR2) found by applying correspondence analysis to the data from Steneryd Nature Reserve. The species names have been given obvious abbreviation, and the sites are labeled S1 to S17.

12.6   *Comparison of ordination methods*

Four methods of ordination have been reviewed in this chapter, and it would be nice to be able to state when each should be used. Unfortunately, this cannot be done in an altogether satisfactory way because of the wide variety of different circumstances for which ordination is used. Therefore, all that will be done here is to make some final comments on each of the methods in terms of its utility.

Principal components analysis can be used only when the values for p variables are known for each of the objects being studied. Therefore, this method of analysis cannot be used when only a distance or similarity matrix is available. When variable values are available and the variables are approximately normally distributed, this method is an obvious choice.

When an ordination is required starting with a matrix of distance or similarities between the objects being studied, it is possible to use either principal coordinates analysis or multidimensional scaling. Multidimensional scaling can be metric or nonmetric, and principal coordinates analysis and metric multidimensional scaling should give similar results. The relative advantages of metric and nonmetric multidimensional scaling will depend very much on the circumstances, but, in general, nonmetric scaling can be expected to give a slightly better fit to the data matrix.

Correspondence analysis was developed for situations where the objects of interest are described by measures of the abundance of different characteristics. When this is the case, this method appears to give ordinations that are relatively easy to interpret. It has certainly found favor with ecologists analyzing data on the abundance of different species at different locations.

12.7 Computer programs

Programs for principal components analysis have been discussed in Chapters 6 and 7 and will not be considered further here. The package MVSP (Kovach, 2003) was used for the calculations in Examples 12.1 and 12.2. This program was also used for the principal coordinates analyses of Examples 12.3 and 12.4, including the construction of distance matrices, and for the correspondence analysis of Example 12.7. It includes a considerable number of options for ordination, including detrended correspondence analysis.

MVSP is one of a number of packages that are designed mainly for ordination and related analyses. Others include CANOCO (ter Braak and Smilauer, 2003), those in the book by Ludwig and Reynolds (1988), and PC-ORD (Digisys, 2003). In addition, the standard statistical packages mentioned in the Appendix to this book include some ordination options.

12.8 Further reading

Suggestions for further reading related to principal components analysis and multidimensional scaling are provided in Chapters 6 and 11, and it is unnecessary to repeat these here. For further discussions and more examples of principal coordinates analysis and correspondence analysis, particularly in the context of plant ecology, see the books by Digby and Kempton (1987), Ludwig and Reynolds (1988), and Jongman et al. (1995). For correspondence analysis, the classic reference is by Greenacre (1984). In addition, there is a short book on correspondence analysis by Clausen (1998) and a very comprehensive book on the same topic by Benzecri (1992).

One important technique not covered in this chapter is canonical ordination, where the ordination axes are chosen to represent a set of explanatory variables as much as possible. For example, there might be interest in seeing how the distribution of plant species over a number of sites is related to the temperature and soil characteristics at those sites. Discriminant function analysis is one special case of this type of analysis, but a number of other analyses are also possible. See Jongman et al. (1995) for more details.

12.9   Chapter summary

- Ordination is the process of producing a small number of variables to represent the relationships between a number of objects, usually graphically. Sometimes the word *scaling* is used instead of ordination.
- Many of the methods described in earlier chapters can be used for ordination. An example based on the abundances of plant species at the Steneryd Nature Reserve is used to illustrate ordination through a principal components analysis. A second example is also provided based on the presence and absence of different types of goods in a cemetery at Bannadi in Thailand.
- Principal coordinates analysis is a method for ordination that starts from a matrix of similarities between n objects, in a similar way to multidimensional scaling. Principal coordinates analysis is related to a principal components analysis. The theory of the method is reviewed, and its use is illustrated using the data from the Steneryd Nature Reserve and the Bannadi cemetery.
- The use of multidimensional scaling for ordination is also illustrated using the data from the Steneryd Nature Reserve and the Bannadi cemetery.
- Correspondence analysis is the fourth ordination method that is discussed. This method is particularly favored by plant ecologists. The theory of the method is reviewed, and it is illustrated using the data from the Steneryd Nature Reserve and the Bannadi cemetery.
- Recommendations are made about when to use each of the four ordination methods that are discussed.
- Computer programs for ordination are discussed.
- Suggestions are made about further reading.

Exercise

Table 6.6 shows the values for six measurements taken on each of 25 prehistoric goblets excavated in Thailand. The nature of the measurements is shown in Figure 6.3. Use the various methods discussed in this chapter to produce ordinations of the goblets and see which method appears to produce the most useful result.

References

Benzecri, P.J. (1992), *Correspondence Analysis Handbook*, Marcel Dekker, New York.

Clausen, S.E. (1998), *Applied Correspondence Analysis*, Sage Publications, Thousand Oaks, CA.

Digby, P.G.N. and Kempton, R.A. (1987), *Multivariate Analysis of Ecological Communities*, Chapman and Hall, London.

Digisys (2003), OC-ORD for Windows, Multivariate Analysis of Ecological Data, Ver. 4; available on-line at www.digisys.net.

Fisher, R.A. (1940), The precision of discriminant functions, *Ann. Eugenics*, 10, 422–429.

Greenacre, M.J. (1984), *Theory and Application of Correspondence Analysis*, Academic Press, London.

Hill, M.O. and Gauch, H.G. (1980), Detrended correspondence analysis, an improved ordination technique, *Vegetatio*, 42, 47–58.

Hirschfeld, H.O. (1935), A connection between correlation and contingency, *Proc. Cambridge Philos. Soc.*, 31, 520–524.

Jongman, R.H.G., ter Braak, C.J.F., and van Tongeren, O.F.R. (1995), *Data Analysis in Community and Landscape Ecology*, Cambridge University Press, Cambridge.

Kovach, W.L. (2003), MVSP — Multi-Variate Statistical Package, Ver. 3.1, Kovach Computing Services; available on-line at www.kovcomp.co.uk.

Ludwig, J.A. and Reynolds, J.F. (1988), *Statistical Ecology*, Wiley, New York.

ter Braak, C.J.F. and Smilauer, P. (2003), CANOCO — a FORTRAN Program for Canonical Community Ordination by (Partial) (Detrended) (Canonical) Correspondence Analysis, Principal Components Analysis and Redundancy Analysis, Ver. 4.5, Plant Research International; available on-line at www.plant.dlo.nl.

chapter thirteen

Epilogue

13.1 *The next step*

In writing this book, my aims have purposely been limited. These aims will have been achieved if someone who has read the previous chapters carefully has a fair idea of what can and what cannot be achieved by the multivariate statistical methods that are most widely used. My hope is that the book will help many people take the first step in "a journey of a thousand miles."

For those who have taken this first step, the way to proceed further is to gain experience of multivariate methods by analyzing different sets of data and seeing what results are obtained. As in other areas of applied statistics, competence in multivariate analysis requires practice.

Recent developments in multivariate analysis have been made in the closely related field of data mining, which is concerned with extracting information from very large data sets. This topic has not been considered in this book, but it is an area that should be investigated by anyone dealing with large multivariate data sets. More details will be found in the book by Hand et al. (2001).

13.2 *Some general reminders*

In developing expertise and familiarity with multivariate analyses, there are a few general points that are worth keeping in mind. Actually, these points are just as relevant to univariate analyses. However, they are still worth emphasizing in the multivariate context.

First, it should be remembered that there are often alternative ways of approaching the analysis of a particular set of data, none of which is necessarily the best. Indeed, several types of analysis may well be carried out to investigate different aspects of the same data. For example, the body measurements of female sparrows given in Table 1.1 can be analyzed by principal components analysis or factor analysis to investigate the dimensions behind body-size variation, by discriminant analysis to contrast survivors and nonsurvivors, by cluster analysis or multidimensional scaling to see how the birds group together, and so on.

Second, use common sense. Before embarking on an analysis, consider whether it can possibly answer the questions of interest. Many statistical analyses are carried out because the data are of the right form, irrespective of what light the analyses can throw on a question. At some time or another, most users of statistics find themselves sitting in front of a large pile of computer output with the realization that it tells them nothing that they really want to know.

Third, multivariate analysis does not always work in terms of producing a neat answer. There is an obvious bias in statistical textbooks and articles toward examples where results are straightforward and conclusions are clear. In real life, this does not happen quite so often. Do not be surprised if multivariate analyses fail to give satisfactory results on the data that you are really interested in! It may well be that the data have a message to give, but the message cannot be read using the somewhat simple models that standard analyses are based on. For example, it may be that variation in a multivariate set of data can be completely described by two or three underlying factors. However, these may not show up in a principal components analysis or a factor analysis because the relationship between the observed variables and the factors is not a simple linear one.

Finally, there is always the possibility that an analysis is dominated by one or two rather extreme observations. These outliers can sometimes be found by simply scanning the data by eye, or by considering frequency tables for the distributions of individual variables. In some cases, a more sophisticated multivariate method may be required. For example, a large Mahalanobis distance from an observation to the mean of all observations is one indication of a multivariate outlier (see Section 5.3), although the truth may just be that the data are not approximately multivariate normally distributed.

It may be difficult to decide what to do about an outlier. If it is due to a recording error or some other definite mistake, then it is fair enough to exclude it from the analysis. However, if the observation is a genuine value, then this is not valid. Appropriate action then depends on the particular circumstances. See Barnett and Lewis (1994) for a detailed discussion of possible approaches to the problem.

Sometimes an effective approach is to do an analysis with and without the extreme values. If the conclusions are the same, then there is no real problem. It is only if the conclusions depend strongly on the extreme values that they need to be dealt with more carefully.

13.3   *Missing values*

Missing values can cause more problems with multivariate data than with univariate data. The trouble is that when there are many variables being measured on each individual, it is often the case that one or two of these variables have missing values. In such cases, individuals with missing values may be excluded from an analysis, resulting in the exclusion of an impractically large proportion of individuals. For example, in studies of ancient human populations, skeletons are frequently broken and incomplete.

Texts on multivariate analysis are often quite silent on the question of missing values. To some extent, this is because dealing with missing values is not a straightforward matter. In practice, computer packages sometimes include a facility for estimating missing values by various methods of varying complexity. One possible approach is to estimate missing values and then analyze the data, including these estimates, as if they were complete data in the first place. It seems reasonable to suppose that this procedure will work satisfactorily, providing that only a small proportion of values are missing.

For a detailed discussion of methods for dealing with missing data, see the recent book by Little and Rubin (2002).

References

Barnett. V. and Lewis, T. (1994), *Outliers in Statistical Data*, 3rd ed., Wiley, New York.

Hand, D., Mannila, H., and Smyth, P. (2001), *Principles of Data Mining*, MIT Press, Cambridge, MA.

Little, R.A. and Rubin, D.B. (2002), *Statistical Analysis with Missing Data*, 2nd ed., Wiley, New York.

appendix

Computer packages for multivariate analyses

The most important consideration with computer packages is that they provide the correct output for the analysis and options selected. Usually it is a fair assumption that the output is correct for very standard analyses, but if unusual options are used, then it is best to study the output carefully to make sure that everything seems correct. It might not be.

Assuming that the accuracy of output is not a consideration, it is clear that the needs in terms of computing for someone just starting to use multivariate methods are not the same as the needs of an expert in this area. The novice needs a computer package that is simple to use, even if this means that the number of options is limited for some analyses. The expert needs many options to be available, even if this means that choosing which options to use requires considerable thought.

In this Appendix, it is the needs of the novice that are considered to be important. What is provided here in Table A1 is a list of some of the many commercial statistical packages that are available, with details about which multivariate analyses they perform, an assessment of how easy each of the packages is for a beginner to use, and also an assessment of the graphics capabilities. Several of the packages listed include a programming language, which makes them much more flexible for the experienced user. However, the assessments provided in the table are only based on analyses that are available using a menu option.

There are a number of specialized packages that are not considered here. For example, CANOCO (ter Braak and Smilauer, 2003) or PC-ORD (Digisys, 2003) may be very suitable for the user who just wants to do ordination. There are also some packages that are add-ins to Microsoft Excel such as XLSTAT-PRO (Xlstat, 2003) that do many of the multivariate analyses described in this book. Such a package may be very suitable for those who are accustomed to using Excel to do all their calculations.

Table A1 Some Statistical Packages with Information about the Multivariate Analyses that They Provide, an Assessment of How Easy the Packages Are to Use for Someone New to Multivariate Analysis, and an Indication of the Graphical Facilities in the Package

Package	Ease of Use[a]	Source of Information	Graphs[b]	Tests of Significance[c]	Distance[d]	Mantel Test	PCA[e]	FA[f]	DFA[g]	CA[h]	Log Reg[i]	CCA[j]	MDS[k]	PCoA[l]	CORA[m]
												Tests and Analyses Available on Menus			
GenStat	*	www.vsn-intl.com	**	Yes	Yes	Yes	Yes	No	Yes	Yes	Yes	Yes	Yes	Yes	Yes
MINITAB 14	**	www.minitab.com	***	Yes	Yes	No	Yes	Yes	Yes	Yes	Yes	No	No	No	Yes
MVSP 3.1	*	www.kovcomp.com	*	No	Yes	No	Yes	No	No	Yes	No	No	No	Yes	Yes
NCSS 2004	**	www.ncss.com	***	Yes	Yes	No	Yes	Yes	Yes	Yes	Yes	Yes	Yes	No	Yes
SPSS 12	**	www.spss.com	***	Yes	Yes	No	No	Yes	Yes	Yes	Yes	Yes	Yes	No	Yes
Stata 8.0	*	www.stata.com	***	Yes	Yes	No	Yes	Yes	No	Yes	Yes	Yes	No	No	No
Statistica 6.1	**	www.statsoft.com	***	Yes	Yes	No	Yes	Yes	Yes	Yes	Yes	Yes	Yes	No	Yes

a Ease of use for beginners: * = reasonably easy after some practice; ** = the easiest.

b Graphics capabilities: * = limited; ** = reasonably good; *** = best.

c Tests of significance as discussed in Chapter 4, although none of the packages have all of these tests as menu items.

d Calculation of distances as discussed in Chapters 5 and 9. Generally only some of these distances are available.

e Principal components analysis.

f Factor analysis.

g Discriminant function analysis.

h Cluster analysis.

i Logistic regression.

j Canonical correlation analysis.

k Multidimensional scaling.

l Principal coordinates analysis.

m Correspondence analysis.

References

Digisys (2003), OC-ORD for Windows, Multivariate Analysis of Ecological Data, Ver. 4.; available on-line at www.digisys.net.

ter Braak, C.J.F. and Smilauer, P. (2003), CANOCO — a FORTRAN Program for Canonical Community Ordination by (Partial) (Detrended) (Canonical) Correspondence Analysis, Principal Components Analysis and Redundancy Analysis, Ver. 4.5, Plant Research International; available on-line at www.plant.dlo.nl.

Xlstat (2003), XLSTAT-PRO, add-in for Excel; available on-line at www.xlstat.com.

Author Index

Barnett, V. 202, 203
Bartlett, M.S. 146, 150, 161
Benzecri, P.J. 191, 197, 198
Bernstein, I.H. 102, 103
Borg, I. 174, 175
Bumpus, H. 1, 3, 16, 27, 28, 29, 38, 39, 43, 45, 86, 118

Carter, E.M. 36, 38, 57
Chatfield, C. 101, 102, 103
Chernoff, H. 30, 31, 34
Clausen, S.E. 197, 198
Cleveland, W.S. 33, 34
Collett, D. 123, 124
Collins, A.J. 101, 102, 103
Cox, M.A.A. 174, 175
Cox, T.F. 174, 175

Darwin, C. 1
Digby, P.G.N. 137, 141, 197, 198
Dunteman, G.H. 85, 90
Digisys 197, 198, 205, 207

Ehrlich, P.R. 16
Everitt, B. 30, 34, 135, 141

Fidell, L.S. 102, 104
Fisher, R.A. 107, 123, 124, 191, 199
Francis, R.I.C.C. 37, 58

Galton, F. 3
Garbin, C.P. 103
Gauch, H.G. 195, 199
Giffins, R. 159, 161
Gower, J.C. 68, 73, 74
Green, E.L. 151, 161
Greenacre, M.J. 197, 199
Groenen, P. 174, 175

Hand, D. 201, 203
Harris, R.J. 109, 124, 147, 161

Hartigan, J. 135, 141
Harvey, H.H. 74
Harville, D.A. 25, 26
Healy, M.J.R. 25, 26
Higham, C.F.W. 10, 16, 55, 57, 62, 74, 87, 137, 179
Hill, M.O. 195, 199
Hintze, J. 101, 104, 122, 124, 131, 141, 167, 170, 172, 175
Hirschfeld, H.O. 191, 199
Hosmer, D.W. 123, 124
Hotelling, H. 3, 37, 38, 41, 44, 49, 54, 75, 90, 143, 161

Jackson, D.A. 68, 73, 74
Jackson, J.E. 85, 90
Jacoby, W.G. 34
Jadwiszczak, P. 72, 74
Jolliffe, I.T. 85, 90
Jongman, R.H.G. 197, 199

Kaiser, H.F. 94, 104
Kempton, R.A. 137, 141, 197, 198
Khatri, C.G. 57
Kijngam, A. 16, 57, 74
Kovach, W.L. 197, 199
Kres, H. 48, 57
Kruskal, J.B. 164, 165, 166, 174, 175

Landau, S. 141
Leese, M. 141
Legendre, P. 68, 74
Lemeshow, S. 123, 124
Levene, H. 42, 44, 53, 54, 57, 82
Lewis, T. 202, 203
Little, R.A. 203
Ludwig, J.A. 190, 197, 199

Mahalanobis, P.C. 63, 64, 65, 66, 67, 72, 73, 74, 105, 106, 109, 111, 114, 116, 123, 202

Manly, B.F.J. 16, 37, 42, 57, 58, 73, 74, 116, 124
Mannila, H. 203
Mantel, N. 69, 70, 71, 72, 73, 74
McKechnie, S.W. 7, 16
Mielke, P.W. 71, 74

Namboodiri, K. 25, 26

Pearson, K. 75, 90
Penrose, L.W. 63, 64, 65, 66, 71, 72, 73, 74
Peres-Neto, P.R. 42, 58, 73, 74
Persson, S. 137, 141

Randall-Maciver, R. 5, 16
Rencher, A.C. 102, 104
Reynolds, J.F. 190, 197, 199
Romesburg, H.C. 135, 141, 169, 171, 175
Rubin, D.B. 203

Schultz, B. 42, 58
Searle, S.R. 25, 26
Seber, G.A.F. 49, 58, 101, 102, 104, 120, 122, 124
Smilauer, P. 197, 199, 205, 207
Smyth, P. 203
Somers, K.M. 74
Spearman, C. 91, 92, 102, 104

Srivastava, M.S. 57
Steyn, A.G.W. 34
Stumf, R.H. 34

Tabachnick, B.G. 102, 104
Teng, G.C. 103
Ter Braak, C.J.F. 197, 199, 205, 207
Thompson, B. 159, 161
Thomson, A. 5, 16
Togerson, W.S. 164, 175
Toit S.H.C. 31, 34
Tuft, E.R. 33, 34

Van Tongeren, O.F.R. 199
Van Valen, L. 43, 44, 51, 54, 58

Weber, A. 88, 89, 90
Welch, B.L. 36, 54, 58
Welsch, R.E. 30, 34
White, R.R. 16
Wish, M. 166, 174, 175

Yao, Y. 38, 58

Xlstat 205, 207

Subject Index

Analysis of variance 46, 51, 54, 107–108

Between-sample matrix of sums of squares and cross-products 47, 108, 110
Bonferroni adjustment, *see* Test of significance

Canonical correlation analysis 13, 143–161, 177
   canonical correlations 143–146, 150, 152
   canonical variables 143–146, 150–151, 152, 156–157
   computer programs 158
   interpreting canonical variates 148, 150–151, 157
   numerical procedure 145–146
   tests of significance 146–147, 150, 157–158
Canonical ordination 197
Chernoff faces 30–31, 33
Chi-squared test, *see* Test of significance
Cluster analysis 13, 72, 125–141, 179, 201
   based on assuming a mixture of several populations 135
   computer programs 134–135
   dendrogram 128
   divisive hierarchic clustering 128
   furthest neighbour clustering 128
   group average clustering 128
   hierarchic methods 125–126, 131, 134
   k-means clustering 131–133
   measures of distance 129–130
   nearest neighbour clustering 127–128, 131, 134
   partitioning methods 126–127
   problems with cluster analysis 129
   uses for cluster analysis 125
   with a principal components analysis 130

Computer programs 15, 54, 72, 84–85, 101, 122, 134–135, 203, 205–207
   CANOCO computer package 197, 205
   GenStat 206
   Minitab 206
   MVSP 197, 206
   NCSS 101, 122, 131, 167, 170, 172, 206
   NMDS 190
   PC-ORD 197, 205
   SPSS 206
   Stata 206
   Statistica 206
   XLSTAT-PRO 205
Correlation coefficient 3, 24–25
Correlation matrix 24–25, 79, 83, 97, 149–150, 178
Covariance matrix 15, 23–25, 37–40, 50, 63, 64–66, 78–79, 109, 122, 184
Correspondence analysis 14, 191–196, 197
   arch or horseshoe plot 195
   computer programs 197
   detrended correspondence analysis 195
   form of data 191–192
   numerical procedure 193–195
   reciprocal averaging 193

Data mining 201
Dendrogram, *see* Cluster analysis
Dice-Sorensen index 68
Discriminant function analysis 13, 105–124, 177, 197, 201
   assigning ungrouped individuals to groups 116
   by logistic regression, *see* Logistic regression
   canonical discriminant functions 107–108, 110–111

discrimination based on
   Mahalanobis distances
   105–106, 111, 114, 116
form of data 105–106
jackknife classification of cases 116
prior probabilities of group
   membership 114
quadratic discriminant functions 122
robustness 109
randomization analysis 116
stepwise 114–116
tests of significance 108–109, 112
Dispersion matrix, *see* Covariance
   matrix
Distance matrix, *see* Multivariate
   distances
Draftsman's plot 29–30, 33, 157–158, 178,
   181, 183, 188, 189, 190, 191,
   192

Eigenvalues and vectors, *see* Matrix
Euclidean distances, *see* Multivariate
   distances
Examples
   boys in a preparatory school 91–92
   colonies of a butterfly 6–8, 59, 74, 143,
      148–151, 158
   Egyptian skulls 3–6, 51–53, 71–72,
      105, 110–111, 116, 120–122
   employment in European countries
      9–11, 83–84, 97–100, 112–114,
      130–133, 159–160, 175, 177
   grave goods from the Bannadi
      cemetery in Thailand 137,
      139–140, 179–183, 190–192
   plant species in Steneryd Nature
      Reserve 137–138, 178–179,
      187–189, 190–191, 195–196
   prehistoric dogs from Thailand 9,
      30–33, 55–57, 59, 61–62, 124,
      133–134
   prehistoric pottery goblets for
      Thailand 87–88 198
   protein consumption in Europe
      88–89, 103, 159–160
   reading and arithmetic tests for
      schoolchildren 143–144
   road distances between New
      Zealand towns 166–169, 177
   soil and vegetation in Belize 151–158

storm survival of sparrows 1–3,
   27–29, 29–30, 37, 38–40, 43–45,
   76, 79–82, 118–119, 177, 201
voting behaviour of New Jersey
   Congressmen 169–174, 177

F-test, *see* Test of significance
Factor analysis 12–13, 91–104, 201, 202
   common factors 92–93
   communality 93, 97
   computer programs 101
   exploratory and confirmatory factor
      analysis 102
   factor loading 92–93, 98
   factor rotation 94–95, 96, 98
   factor scores 94, 96–97
   form of data 93
   Kaiser normalization 94, 98
   maximum likelihood 101
   number of factors 94, 97, 101
   specificity 93
   structural equation modelling 102
   two factor theory of mental tests 92
   Value of factor analysis 102
   varimax rotation 94, 96, 98

Graphical methods 15, 27–34

Hotelling's T^2 test, *see* Test of
   significance

Index variable plots 27–29, 33

Jaccard index 68

Levene's test, *see* test of significance
Logistic regression
   generalization to polytomous
      regression 123
   maximum likelihood 117
   sampling schemes 117–118
   separate sampling 118, 120
   tests of significance 119, 121
   use for discrimination 117–122

Mahalanobis distance, *see* Multivariate
   distances
Mantel's matrix randomization test
   69–73, 74

Matrix
    addition 19
    column vector 17–18
    correlation 69, 71
    determinant 21, 46–47, 52
    diagonal 18
    eigenvalues and eigenvectors 22–23,
            47, 48, 49, 53, 78–80, 83, 94,
            97–98, 108, 110–111, 113,
            145–146, 150, 178–179, 180,
            182, 183–184, 186–187, 188,
            193–194, 195
    equal 19
    identity 18
    inverse 21, 40, 63
    latent roots and vectors, *see*
            eigenvalues and eigenvectors
    multiplication 19, 20
    orthogonal 22
    quadratic form 22
    row vector 18
    scalar 20
    singular 22
    square 17
    subtraction 19
    symmetric 18, 184
    trace 19, 78
    transpose 18
    zero 18
Mean vector 15, 23–24, 38, 39, 63
Missing values 202–203
Mixtures of distributions, *see* Cluster
    analysis
Multidimensional scaling 13–14,
        163–175, 177, 181–182, 181,
        201, 202
    choice of the number of dimensions
        166
    computer programs 172–173, 190, 197
    disparities 165, 172, 174
    for ordination 189–192, 196–197
    goodness of fit (stress) 165–166, 170,
        183, 190
    interpretation of dimensions 170–172
    metric scaling 166, 170
    monotonic regression 165, 167
    nonmetric scaling 166, 167
    numerical procedure for a classical
        multidimensional scaling
        165–166

    on presence and absence data
        190–192
Multiple regression 144
Multiple testing, *see* Test of significance
Multivariate analysis of variance
    (MANOVA) 54
Multivariate distances
    between individuals 59–62
    between populations and samples
        59, 62–67
    Euclidean 60–62, 130, 131, 165,
        184–185, 190, 191, 192
    for presence-absence data 68–69, 179,
        187, 190
    from proportions 67
    Mahalanobis 63–67, 72, 105–106, 109,
        111, 114, 116, 202
    Manhattan 187
    niche overlap 67
    Penrose 63–67, 71–72
    similarity indices 67, 186, 187–188
    with cluster analysis 129–130
    with multidimensional scaling
        163–165
Multivariate normal distribution, *see*
    Normal distribution

Niche overlap, *see* Multivariate
    distances
Normal distribution 14–15, 36–37, 38, 42,
    49, 53, 54, 64, 71, 109, 122, 135,
    152, 202

Ochiai index 68
Ordination 14, 27, 72, 177–199
    comparison of ordination methods
        196–197
    computer programs 197
    using correspondence analysis, *see*
        Correspondence analysis
    using multidimensional scaling 177,
        189–192
    using principal components analysis
        177, 178–181
    using principal coordinates analysis,
        see Principal coordinates
        analysis
Outliers, *see* Residuals and outliers

Penrose distance, *see* Multivariate
　　distances
Polytomous regression 123
Presence-absence data, *see* Multivariate
　　distances
Principal components analysis 3, 12, 13,
　　14, 75–90, 91, 93, 95, 97, 143,
　　177, 201, 202
　　computer programs 84–85, 197
　　factor analysis via principal
　　　　components 95–100
　　for ordination 177–181, 196
　　form of data 76–77
　　number of principal components
　　　　used for analyses 79, 80, 83
　　numerical procedure 76–79
　　plots of principal components 82,
　　　　84–85
　　relationship to principal coordinates
　　　　analysis 183–187
　　with cluster analysis 130
Principal coordinates analysis 14, 164,
　　178, 181–189, 196, 197
　　computer programs 187, 197
　　double centring of similarity matrix
　　　　186
　　numerical procedure 186–187
　　relationship to principal components
　　　　analysis 183–187
　　with presence and absence data
　　　　187–189
Procrustes analysis 73
Profile plots 32–33

Reciprocal averaging, *see*
　　　　Correspondence analysis
Residuals and outliers 64, 109, 202

Scaling, *see* Ordination
Similarity measures, *see* Multivariate
　　distances
Simple matching index 68
Size and shape 76, 80–81, 88
Spatial correlation 73, 74, 157
Star plots 30–31, 33

Test of significance
　　Bartlett's test for significant canonical
　　　　correlations 146–147, 150, 152
　　Bonferroni adjustment 41–42
　　Box's M-test 42, 49–51, 53
　　chi-squared test 64, 108–109, 119, 121,
　　　　147, 150
　　comparison of mean values 35–40,
　　　　46–49
　　comparison of variances 42–46, 49–51
　　F-test 38, 40, 42, 46, 47, 48, 50–51, 52,
　　　　54, 82
　　Hotelling's T^2 test 37–40, 41, 42, 43–44
　　Lawes-Hotelling trace test 48–49, 53
　　Levene's test 42, 43–44, 53, 82
　　Pillai's trace test 48–49, 53
　　robustness 36–37, 38, 42, 49, 51, 53, 54
　　Roy's largest root test 47–48, 53
　　multiple tests 35, 41–42
　　tables of critical values for
　　　　multivariate tests 48
　　t-test 36–40, 42, 44, 82
　　Van Valen's test 43–44, 51
　　Welch's test 36–37
　　Wilk's lambda test 46–48, 112
　　with canonical correlation analysis
　　　　146–147
　　with discriminant function analysis
　　　　108–109
Total sum of squares and cross-products
　　matrix 46, 51–52, 108, 110
Two factor theory of mental tests, *see*
　　Factor analysis

Van-Valen's test, *see* Test of
　　significance

Within-sample sum of squares and
　　cross-products matrix 46, 52,
　　108, 110